The AI-Powered Bank

How Machine Learning Is Reshaping Lending, Fraud, and Customer Experience

Joseph Trek

Published by NobleTrex Press

For permissions and other inquiries, write to:

P.O. Box 3132, Framingham, MA 01701, USA

Contents

Introduction **1**

1 Foundations of Modern Banking and AI **9**

1.1 The Business of Banking 9

1.2 Demystifying Machine Learning 19

1.3 Evaluating and Interpreting AI 29

1.4 The Rise of Generative AI in Banking 38

2 The Engine Room: Data Platforms and Trust **49**

2.1 Data Governance and Lineage: The Farm-to-Table Approach 49

2.2 Privacy, Consent, and Security: Keeping Secrets Secret 61

2.3 Integration with Core Systems: The Clash of Generations 71

2.4 Features and Data Leakage: How Models Cheat . . . 80

3 Revolutionising Lending and Credit Risk **95**

3.1 From Traditional Scoring to ML Underwriting 95

3.2 Expanding Access with Alternative Data 105

3.3 Fair Lending and Adverse Action 114

3.4 Pricing, Collections, and Stress Testing 121

4 The Front Lines: Fraud, AML, and Financial Crime **129**

4.1 The Cat and Mouse Game: Fraud vs. Money Laundering 129

4.2 The 300-Millisecond Decision 138
4.3 Uncovering Hidden Networks 147
4.4 The Human Feedback Loop 154

5 Reimagining Customer Experience and Growth 163

5.1 Conversational AI and Contact Centers 163
5.2 Personalisation and Recommendation Engines 172
5.3 Voice and Text Analytics 180
5.4 Value Measurement and ROI 188

6 Governing and Deploying AI Confidently 201

6.1 Building the Guardrails: Model Risk Management . . 201
6.2 Navigating the Rules: Regulation and Vendor Risk . . 211
6.3 Keeping the Lights On: MLOps and Continuous Monitoring 219
6.4 The Human Element: Operating Models and the Road Ahead 227

Introduction

Consider the last time you tapped your card or phone to pay for something. Maybe it was a coffee, a train ticket, or a late-night grocery run. You heard a beep, saw a green checkmark, and walked away. The entire interaction took less than half a second.

In that fleeting blink of time, a massive, invisible machinery roared to life. Your request traveled through a labyrinth of legacy cables, bounced off a satellite or fiber optic line, arrived at a data center, and asked a question: *Should we allow this?*

For decades, the answer to that question was determined by a rigid set of rules written by humans. If the balance is greater than X, and the location is Y, then approve. It was a digital fortress of "if-then" statements-sturdy, predictable, and profoundly dumb.

Today, that fortress is being dismantled. When you tapped your card just now, you weren't judged by a rulebook. You were analyzed by a brain.

A probability engine, trained on billions of historical transactions, compared your coffee purchase against everything it knows about you-and people *like* you. It looked at the time of day, the way you held your phone, the geolocation of your device relative to the merchant, and thousands of other subtle variables that a human analyst couldn't correlate in a lifetime. It decided, with 99.9% confidence, that you are who you say you are. And it did it in 200 milliseconds.

This is not science fiction, nor is it a forecast for the next decade. This is the operating reality of modern finance. We have handed the keys to the vault over to Machine Learning.

The implications of this shift are staggering, terrifying, and exhilarating in equal measure. We are moving from a financial system based on bureaucratic rigidity to one based on fluid, probabilistic intelligence. This book is about that transition. It is about what happens when the oldest business model in the world-banking-collides with the most disruptive technology of our species.

The Death of the Average

To understand why this matters, you have to appreciate how blunt the tools of banking used to be. For the last century, banking was the business of averages.

If you applied for a loan, you weren't really an individual; you were a credit score. You were a number between 300 and 850. If you were a 720, you were "good." If you were a 620, you were "risky." It didn't matter if you were a responsible gig-economy worker who just happened to lack a long credit history; the rules said you were a risk, so the bank said no.

This "law of averages" was necessary because banks are businesses of scale. A human loan officer can't deeply interview ten million customers. They needed shortcuts. They needed broad buckets.

Machine learning kills the average. By ingesting vast amounts of alternative data-cash flow patterns, utility payments, even the stability of your digital identity-AI allows banks to see the individual grain of the wood rather than just the forest. It permits a level of granularity that was previously impossible.

For the single mother with a thin credit file but impeccable cash management, this is a revolution. It means access. It means a mortgage is suddenly on the table because the algorithm sees her *behavior*, not just her history.

But there is a shadow side to this granularity. If an algorithm can predict creditworthiness with uncanny accuracy, it can also predict profitability. It knows exactly how much to charge you to maximize the bank's return without causing you to default. It can optimize fees based on your psychological tolerance. When the bank knows you better than you know yourself, the balance of power shifts dramatically.

The Arms Race in the Shadows

While marketing departments love to talk about AI-driven "financial wellness" and "hyper-personalization," the real money is being spent in the trenches of a silent war.

Every second of every day, banks are under attack. Organized crime syndicates, state-sponsored hacking groups, and lone wolves are testing the fences. In the old days, fraud prevention was a game of whack-a-mole. Fraudsters would find a loophole; banks would write a new rule to close it.

That doesn't work anymore. The bad guys are using AI, too. They are using generative models to create synthetic identities-fake people who look real, act real, and build up credit scores before busting out with millions of dollars in loans. They are using deepfakes to bypass voice authentication. They are automating attacks at a scale that human teams cannot comprehend.

The only defense against an AI that attacks is an AI that defends. This book dives deep into that adversarial loop. We are looking at a future where neural networks duel with one another in real-time, escalating in complexity, with your savings account sitting in the crossfire. The banks that survive this decade will not be the ones with the nicest branches; they will be the ones with the best defensive algorithms.

The Black Box Problem

As we hand over decision-making authority to these mathematical models, we run headlong into a philosophical and legal crisis. It is the "Black Box" problem.

Imagine you apply for a small business loan to expand your bakery. The AI denies you. You ask the bank manager why. In 1990, the manager would pull out a sheet of paper and say, "Your debt-to-income ratio is too high."

Today, the manager might look at their screen, see a red "DENIED" flag, and have absolutely no idea why it's there. The model made a decision based on the interaction of five hundred variables in a non-linear high-dimensional space. It found a pattern of risk that human language lacks the words to describe.

Is that acceptable? In a regulated society, can we allow a machine to affect our livelihoods if it cannot explain itself?

Regulators are currently scrambling to draw lines in the sand. They are demanding "explainability"-the ability to reverse-engineer the brain of the machine. But there is a trade-off. Often, the most accurate models are the most opaque, and the most explainable models are the least accurate. Banks are forced to choose: do we want to be right, or do we want to be understood?

Data: The New Uranium

None of this magic works without fuel. In the age of AI, data is not just an asset; it is the physics of the business. But here is the dirty secret of banking: the data is a mess.

Most banks are running on technology stacks that date back to the Cold War. They have customer data siloed in different departments-credit cards on one mainframe, mortgages on another, savings on a third. To build an "AI-Powered Bank," you cannot just sprinkle some machine learning dust on top of a 1970s COBOL core.

We will explore the immense, unglamorous engineering challenge of plumbing. How do you take petabytes of sensitive, regulated financial data and make it usable for modern algorithms without leaking it to hackers or violating privacy laws? This is where the dreams of AI usually die. It is easy to build a model in a lab; it is incredibly hard to deploy it in a bank that has been running for 150 years and relies on code written by people who have long since retired.

Furthermore, data is radioactive. Use it well, and it powers cities. Mishandle it, and it contaminates everything. We have to talk about bias. If a bank trains its lending AI on fifty years of historical loan data, and that historical data reflects fifty years of human prejudice, the AI will not be neutral. It will be a mathematical amplifier of racism and sexism. Cleansing the data of our past sins is one of the hardest technical and ethical challenges discussed in these pages.

The Human Element

You might expect a book about AI to be primarily about code, tensors, and cloud computing. But at its heart, this is a book about management, culture, and people.

The biggest barrier to the AI-powered bank is not technology; it is trust. How does a CEO trust a model that predicts a recession when the economists disagree? How does a compliance officer trust a chatbot to talk to customers without hallucinating a new refund policy? How does a customer trust a bank that watches their spending habits with the eye of a surveillance state?

We are moving toward a world of "Self-Driving Money." In the near future, you won't manually move money into savings or decide which debt to pay off first. An autonomous agent, acting on your behalf, will constantly sweep your liquidity to the place where it earns the highest yield or reduces the most interest. It will be like having a private CFO in your pocket.

But for that to work, the bank has to be more than a vault. It has to be a technology company that happens to have a banking license.

How to Read This Book

We are about to dissect the anatomy of this new organism.

We begin in the engine room. We have to understand the business model of banking-how banks actually make money-because AI is expensive, and if it doesn't drive the bottom line, it's just a toy. We will strip away the hype and define what machine learning actually is in this context, separating the practical tools from the marketing buzzwords.

From there, we look at the raw materials. We will examine the data infrastructure and the governance required to keep it safe. If you are a practitioner, this is your blueprint for avoiding disaster. If you are a leader, this is your guide to asking the right questions before signing the check for a new data platform.

Then, we go to the front lines of the business: Lending and Credit. We will see how underwriting is transforming from a static snapshot to a moving video. We will grapple with the hard math of fairness and the economics of risk-based pricing.

We will pivot to the shield: Fraud and Financial Crime. You will learn how graph analytics can spot a money-laundering ring by analyzing the shape of the network, rather than the content of the transaction. We will look at the speed of the decision-how to stop a thief in the 300 milliseconds before the payment clears.

We will then turn to the face of the bank: Customer Experience. This is where the rubber meets the road for the average user. We will look at how Large Language Models (LLMs) are revolutionizing the call center and how banks are trying to cross the "Uncanny Valley" to provide service that feels personal, not robotic.

Finally, we end with the guardrails. Governance is not the boring part of the story; it is the only thing keeping the whole system from collapsing. We will explore the emerging regulatory landscape, from the EU AI Act to US fair lending guidance, and discuss how to manage the risk of models that drift, decay, or go rogue.

A Note on Optimism

It is easy to be cynical about banks. They are easy villains. And it is easy to be fearful of AI. It is an alien intelligence.

But when you combine them correctly, there is a tremendous capacity for good. An AI-powered bank can identify a customer in financial distress weeks before they miss a payment and offer a solution rather than a penalty fee. It can spot the early signs of elder abuse in transaction data and alert a family member. It can extend capital to entrepreneurs who have been invisible to the traditional system for generations.

The transition is inevitable. The "dumb" bank is extinct; it just hasn't fallen over yet. The question is not whether banks will adopt AI, but whether they will do it responsibly, effectively, and safely.

The money in your pocket is data. The institution that holds it is a software company. The decisions that shape your financial life are algorithms. It is time we understood how they work.

Chapter 1

Foundations of Modern Banking and AI

Before we entrust our financial future to algorithms, we must understand the machinery they are replacing. This chapter bridges the gap between the centuries-old business of interest spreads and the futuristic world of neural networks. We will strip away the jargon to reveal how banks actually make money, why risk is their primary inventory, and how Artificial Intelligence is evolving from a back-office utility into the decision-making brain of the modern financial institution.

1.1 The Business of Banking

A bank looks like a place where money *sits*. In reality, money in a bank is almost always *in motion*. Depositors park cash for safety and convenience; borrowers pull cash forward in time to buy a home, fund payroll, or smooth a rough season. The bank is the matchmaker—and it earns its living in the gap between what it pays for money and what it charges for money. That gap has an unglamorous name: *net interest margin* (NIM). It is also the oxygen of traditional banking.

Net interest margin is the spread between interest a bank *earns* on assets (mostly loans, sometimes securities) and interest it *pays* on liabilities (mostly deposits, sometimes wholesale borrowing), scaled to the size of the balance sheet. Think of it as the bank's "gross profit margin" for the basic act of moving money from savers to borrowers.

A typical simplified story sounds like this: depositors receive 2%, borrowers pay 6%, and the bank has 4% to run the machine and absorb the inevitable dents. The reality has more gears, but the logic holds: banking is a spread business wrapped around risk.

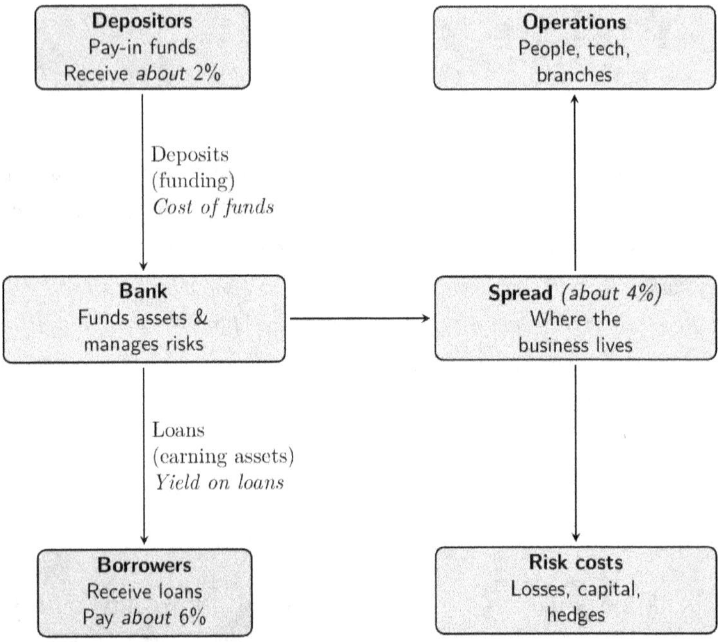

The diagram is deliberately simple—and that simplicity is useful. If you can keep this picture in your head, most bank strategy debates become less mysterious. Why are executives obsessed with "deposit franchise"? Because deposits are often the cheapest, stickiest funding a bank can get, and cheaper funding widens the spread. Why does everyone care about underwriting? Because credit losses punch holes in that spread. Why do banks talk like meteorologists about interest rates? Because rate changes rewrite the economics of both sides of the spread, often at different speeds.

The spread is not a fee for "moving money." It is the price customers collectively pay for a bundle of services and protections that are easy to take for granted: payments, cash access, fraud controls, account statements, customer support, compliance, and the quiet assurance that your checking account won't vanish because your neighbor's business defaulted. The spread is the revenue stream that pays for the plumbing—and for the safety rails.

A bank is a balance sheet with a nervous system

Banks don't sell a single product in the way a retailer sells shoes. They run a balance sheet: a structured pile of assets funded by liabilities, with a thin slice of equity at the bottom absorbing shocks. The classic core looks like this:

Assets (what the bank owns)	*Liabilities & Equity (how it's funded)*
Loans (mortgages, cards, small business, auto)	Deposits (checking, savings, CDs)
Securities (Treasuries, MBS, municipals)	Borrowings (wholesale funding, repos)
Cash & reserves	Equity (shareholders' capital)

Nothing in that table is exotic. The twist is the business logic: the bank funds long-lived, risky assets (loans) with liabilities that can be short-lived and emotionally fragile (deposits can leave). This is called *maturity transformation*: borrowing short and lending long. It's economically valuable—society wants savers to have liquidity and borrowers to have long-term funding—but it means the bank is constantly managing timing mismatches, rate mismatches, and human behavior.

That "nervous system" is risk management. It is not a compliance appendix. It is the product team for the bank's actual product: safely extending credit at scale.

The spread isn't profit; it's raw material

The 4% in the simple story is not the bank's take-home pay. It's more like flour in a bakery: necessary, not sufficient. Before a bank earns a dollar of profit, the spread gets consumed by several hungry line items.

Operations eats first. Running a bank means paying for payment rails, core banking systems, customer service, branch leases (if you have them), fraud monitoring, cybersecurity, card networks, call centers, audits, and the unglamorous work of mailing forms and keeping records. Modern banking has the feel of software, but it carries the cost structure of a regulated utility.

Credit losses eat next—and they eat unevenly. Some loans pay like clockwork and then suddenly don't. Losses arrive in clusters (recessions, regional downturns, industry-specific shocks), and they're rarely polite enough to match last year's averages. The bank tries to price for expected losses, but the real business challenge is surviving the *unexpected* losses: the tail risk that shows up right when everyone is least in the mood for it.

Liquidity and interest-rate risk also take bites, often invisibly. A bank may choose to hold more cash-like assets than pure profit-maximization would suggest because it needs confidence it can meet outflows. It may hedge interest-rate exposure, paying for derivatives or accepting lower yield to avoid a balance-sheet surprise. These choices reduce today's margin in exchange for tomorrow's survival.

Finally, *capital* takes its share. Banks are required (by regulation and by common sense) to fund themselves with a minimum amount of equity. Equity is expensive relative to deposits because shareholders demand a higher return for taking the first loss. Holding more capital makes a bank safer but pushes down return on equity unless the bank can earn enough spread to compensate.

So when you hear "NIM expanded 20 basis points," it can sound like a victory lap. But NIM is only one dial. A bank can widen margin by reaching for risk (making riskier loans at higher rates) and then give it all back in charge-offs. The bank that wins is the one that can produce margin *and* keep the loss line boring.

Risk is the core product, not the side effect

Many industries treat risk as a nuisance: something to insure away or litigate away. Banking is different. Risk is the raw ingredient. A loan is literally a bet about the future: that the borrower will have the willingness and ability to repay over time, under changing conditions, with life happening in the background. The bank's job is to place millions of these small bets such that the portfolio behaves predictably enough to fund itself, satisfy regulators, and still earn a return.

That sounds abstract until you notice how banks talk internally. They don't say, "We sold $500 million of mortgages." They say, "We originated $500 million with an average FICO of X, LTV of Y, and DTI of Z." Translation: we took on a particular shape of risk. The money amount is the headline; the *distribution* is the story.

A useful mental flip is this: banks are not paid for lending money; they are paid for *absorbing uncertainty on behalf of depositors*. Depositors want certainty—my money is there when I need it. Borrowers live with uncertainty—my income could drop, my expenses could spike, my customer could delay payment. The bank stands in between and, for a price, smooths the uncertainty into something that looks stable.

Do this well and the bank earns a steady spread. Do it poorly and the spread becomes an illusion—yesterday's income statement hiding tomorrow's losses.

Why a ``good" bank still has bad loans

A common misunderstanding is that strong banks "avoid bad loans." They don't. A strong bank expects some loans to fail and builds a business that can absorb that failure. Refusing all risk would mean making almost no loans, which would mean earning almost no spread, which would mean being a very safe nonprofit with a banking license.

The real question is: *How many loans will fail, how badly, and under what conditions?* That's a forecasting problem, not a moral one. Banks try to estimate *probability of default* (how likely a borrower is to miss payments), *loss given default* (how much is lost after recoveries), and *exposure at default* (how big the balance will be when things go wrong). These three ideas quietly run the economics of lending—even when nobody says their names out loud in a marketing meeting.

A bank that prices loans as if defaults don't exist is not "optimistic." It is undercharging for its product. Undercharging creates market share today and insolvency later. This is why risk management isn't just about saying "no." It's about charging correctly for "yes," and making sure the bank can live with the consequences when "yes" occasionally turns into "oops."

The lending lifecycle: where spreads are born and where they die

A loan starts life as a promise and ends life as a stream of payments—or a loss. Between those endpoints is a lifecycle with several points of failure:

Marketing and acquisition. The bank decides which customers to pursue, what offers to make, and what channels to use. Bad targeting can be expensive in a boring way (high acquisition cost) or in a dramatic way (attracting borrowers who take the offer because other lenders said no).

Underwriting. This is where the bank turns a person or business into a risk profile. Underwriting decides whether to approve, how much to lend, at what rate, with what collateral, and with what covenants. Underwrite too tightly and you turn away profitable customers. Underwrite too loosely and you book future losses while celebrating today's growth.

Pricing. Pricing is underwriting with a calculator. Two borrowers can both be "approved" and still be wildly different economic deals. The rate, fees, and structure must cover expected losses, operating costs, funding costs, and the return demanded by shareholders—while still being competitive. Pricing is where the spread becomes real, because it links the borrower's behavior to dollars.

Servicing. The loan then becomes an operations business: billing, payment processing, escrow, customer support, disputes, modifications. Servicing quality can reduce losses (early outreach, payment plans) or create them (errors that push borrowers into delinquency). It's also where customer experience is either earned or squandered.

Collections and recovery. When things go wrong, the bank shifts from customer service to damage control. The skill here is not only legal enforcement; it's deciding when to restructure, when to pursue collateral, and when to stop throwing good money after bad.

In every phase, the spread is either protected or punctured. The bank that treats lending as "approve and forget" is essentially betting that the world will remain static. It won't.

Interest rates: the invisible hand on your income statement

Banks are often described as "rate-sensitive," but that phrase hides a more interesting truth: banks are *timing-sensitive.* Assets and liabilities reprice on different schedules. A credit card portfolio might reprice quickly as rates change; a fixed-rate mortgage doesn't. Deposits might be slow to reprice at first and then suddenly become expensive when customers notice money market alternatives. Securities might lock in yield for years. Meanwhile, borrowers' ability to pay changes with the economy—sometimes because rates changed, sometimes because jobs did.

This is why interest-rate risk can feel like physics. You can't negotiate with it. If a bank funds long-term fixed-rate loans with short-term deposits, then a rapid rise in deposit costs can squeeze margin even if the loan book looks "high quality." The borrower may be paying on time, yet the bank's economics deteriorate because the bank's cost of funding has moved and the asset yield hasn't.

Risk management here is partly math (duration, convexity, stress tests) and partly realism about human behavior. Depositors are not a homogeneous blob. Some balances are "sticky" because they're tied to payroll and bill pay. Others are "rate-chasing" because they're excess cash that can move with a few taps. Liquidity and interest-rate risk management is the art of not mistaking temporary calm for permanent stability.

Liquidity: the bank run is a confidence problem wearing a finance costume

A bank can be solvent and still fail. Solvent means assets exceed liabilities in the long run—your loans are worth more than what you owe depositors. But a bank run is a short-run event: depositors demand cash *now*. Even good loans are hard to turn into cash instantly without selling at a discount. That discount is where liquidity problems become solvency problems.

The "bank run" is often told as a dramatic scene—lines, panic, headlines. The modern version can be quieter: large uninsured balances leaving via wire transfers at machine speed. Either way, the core ingredient is the same: confidence. Depositors don't need to prove the bank is insolvent. They only need to suspect that other depositors might leave first.

This is another reason the spread is not free money. Banks maintain liquidity buffers, diversify funding sources, monitor concentration in deposit bases, and plan for stress. All of that costs margin. But the alternative is discovering, at the worst possible time, that profitability without liquidity is just a well-decorated trapdoor.

Fees and ``noninterest income'': the second engine (and the reason big banks can look different)

Not every dollar a bank earns comes from the spread. Banks also earn *noninterest income*: fees for service, interchange from card spending, investment and wealth management revenue, advisory fees, and more. This matters because it changes how the business behaves.

A bank with strong fee income can tolerate a thinner NIM. It can price loans more competitively, invest more in technology, and still produce acceptable returns. A bank with little fee income is more dependent on spread and therefore more exposed to margin compression when deposit costs rise or competition intensifies.

This difference shapes strategy. Some banks build ecosystems—cards, payments, treasury management for businesses—because those services create both revenue and stickier deposits. Others lean into specialized lending niches where they believe they can price risk better than competitors. Both approaches are attempts to answer the same question: *How do we earn enough spread-like economics without taking reckless risk?*

Why ``pricing risk'' is not the same as ``being pessimistic''

The phrase "pricing risk" can sound like a banker's way of justifying higher rates. But pricing risk is also a discipline that protects good borrowers. If a bank fails to distinguish between low-risk and high-risk customers, the low-risk customers subsidize the high-risk ones through higher average pricing. Over time, the best borrowers leave for cheaper options, and the bank is left with a riskier pool—an adverse selection spiral that ends with painful losses.

Good risk pricing is not about charging everyone more. It's about charging everyone *appropriately*. That requires measurement, and measurement requires data: income, cash flow, collateral values, payment histories, business performance, macro conditions, fraud signals, and the features hidden in behavior (how accounts are used, not just what people claim on forms). When measurement improves, pricing can become both more accurate and, for many customers, more fair—because it reduces the need to rely on blunt proxies.

The bank that can measure risk precisely can do something almost magical: lend profitably to customers who look risky under coarse rules but are actually reliable under a richer view of the world. That is competitive advantage expressed as inclusion, and it's one of the reasons advanced analytics became such a force in lending long before generative AI showed up.

The bank's real inventory is trust

A retailer holds inventory on shelves. A bank holds trust on its balance sheet. Depositors are extending trust that their money is safe and accessible. Borrowers are extending trust that the bank will treat them predictably—process payments correctly, honor terms, and not turn a temporary hardship into a bureaucratic nightmare. Regulators and markets extend trust that reported numbers mean what they say.

Trust is fragile because banks are leveraged institutions: a relatively small slice of equity supports a very large pile of assets. Leverage is not a moral failing; it's the design that allows banks to amplify savings into credit. But it means errors compound. A few percent of unexpected losses can erase years of earnings. When that reality meets public confidence, you get a business where risk management isn't a department. It's the immune system.

That immune system has to work at multiple levels at once. It has to block obvious dangers (fraud, identity theft, synthetic identities). It has to manage chronic conditions (delinquencies, charge-offs, operational errors). It has to survive rare catastrophes (recessions, sudden rate shocks, rapid deposit outflows). And it has to do all of this while still saying "yes" often enough to justify existing.

A practical way to read a bank: follow the spread, then follow what consumes it

If you want to understand a bank quickly, start with the spread and then ask what eats it. How expensive is the funding mix? How sticky are deposits? How risky is the loan book? How much is spent to operate and comply? How much capital is required for the chosen risks? What happens under stress?

Those questions have a useful side effect: they remove the romance from banking without removing the fascination. A bank is not a pile of money; it's a machine that transforms time, liquidity, and uncertainty. The machine can be run conservatively or aggressively, competently or carelessly. And because the machine is built from promises—deposit promises and loan promises—its success depends less on optimism and more on accuracy.

Accuracy is the hidden theme of modern banking. Accuracy about who will repay. Accuracy about what will happen when rates move. Accuracy about which transactions are genuine. Accuracy about which costs will surge under stress. Over the last few decades, the industry has become a competition to be more accurate, more consistently, in more places at once. That competitive pressure is the doorway through which machine learning entered banking: not as a futuristic add-on, but as a practical tool for protecting the spread from the many ways reality tries to take it back.

1.2 Demystifying Machine Learning

Banking already runs on predictions; the only question is whether those predictions live in someone's head, in a policy binder, or inside a model. When a lender says, "We like stable income," that's a prediction that stable income correlates with repayment. When a fraud team says, "This pattern looks wrong," that's a prediction that "wrong-looking" correlates with theft. Machine learning (ML) doesn't introduce prediction into the bank—it industrializes it. The real shift is not that the bank starts guessing; it starts *measuring its guesses*.

The phrase "artificial intelligence" tends to summon either robots or apocalypse. Machine learning is more ordinary and more dangerous than both: it's a way to turn yesterday's outcomes into tomorrow's decisions, at scale, without having to fully explain the rule in words. The bank isn't buying a crystal ball. It's buying a very efficient rule-finder—one that is only as wise as the data it learns from and the incentives wrapped around its use.

A clean way to understand ML is to compare it to traditional programming. In traditional coding, a human writes explicit rules: if income is above $X and debt is below $Y, approve the loan; if a transaction happens in a new country and the amount is above $Z, decline it. The computer is obedient. It does exactly what the rules say, including the parts you didn't mean.

Machine learning flips the workflow. Instead of writing the rules, you provide examples. You give the computer data about customers or transactions and the outcome you care about—paid back or defaulted, fraud or not fraud, churned or stayed—and you ask it to infer rules that map inputs to outputs. In other words, you're not telling the machine what "risky" means. You're letting it build a working definition of "risky" from history.

That sounds suspiciously like "teaching," which is why ML metaphors often borrow from parenting. Imagine trying to teach a child what a cat is. You could attempt a definition: "A small domesticated mammal with fur, retractable claws, and whiskers." Then a neighbor shows up with a hairless Sphynx, and the child reasonably concludes you were lying. Or the child sees a fox and declares it a cat because it has fur and whiskers. Definitions are brittle because the world is messy.

Now imagine the alternative: you show the child many pictures and point. Cat. Cat. Not cat. Cat. Not cat. Over time, the child forms a concept that survives weird lighting, odd angles, and unfamiliar breeds. The child can't necessarily articulate the full rule ("the eigen-whatever of whisker geometry"), but the concept works. That's supervised learning in spirit: examples plus answers create a decision ability.

The reason banks embraced supervised learning is painfully practical: the world produces labels. Loans either become delinquent or they don't. Customers either accept an offer or ignore it. Transactions are either confirmed as fraud or cleared. Wherever you have a reliable historical outcome, you have the raw material for supervised ML.

Traditional coding vs. machine learning: the swap that changes everything

The switch from "rules-first" to "data-first" sounds like a minor operational preference. It is not. It changes who is accountable for what, how mistakes happen, and how failures hide.

With traditional rules, the failure mode is usually obvious. Someone set the threshold wrong. Someone forgot an exception. Someone didn't anticipate a corner case. The error is a sentence in a policy document wearing a different font.

With ML, the failure mode can be uncanny. The model may be "right" on average and still wrong in a way that matters. It may learn a shortcut that looks like intelligence but is actually an exploit of your dataset. It may rely on patterns that existed in one economic regime and evaporate in the next. It may be accurate in aggregate while being unreliable for a subset of customers that you care about for legal, reputational, or ethical reasons.

That doesn't mean ML is irresponsible. It means ML is a different kind of responsibility: you're responsible not only for the logic you intended, but for the logic the model discovered. In banking, where "because the computer said so" is never an acceptable explanation, that difference matters.

Supervised learning: predictions with a scoreboard

Supervised learning is the workhorse of modern banking because it comes with a built-in discipline: a scoreboard. You can test predictions against reality.

In credit risk, supervised models often predict some version of "Will this customer default?"—sometimes as a probability, sometimes as a score, sometimes as a rank compared with other applicants. The exact label varies. One bank may train a model to predict 90+ days past due within 24 months. Another might predict charge-off within 12 months. The business question is the same: what is the likelihood this loan becomes a loss, and how soon?

In fraud, supervised models often predict "Is this transaction fraudulent?" based on transaction metadata (merchant type, amount, time of day), device signals, behavioral patterns, and network features (how this card or device connects to others). The model's job isn't to philosophize about honesty; it's to decide how much friction to apply in the next 300 milliseconds.

In marketing and customer experience, supervised learning predicts "Will this customer respond?" or "Will they churn?" The bank uses that to decide who receives an offer, what channel to use, and how to allocate scarce attention from relationship managers. At its best, this looks like relevance. At its worst, it looks like a machine that learned how to be annoying with exquisite efficiency.

The reason supervised learning is so tempting is that it turns messy judgment into a repeatable pipeline. But the pipeline can produce garbage if you misunderstand three deceptively simple questions: What exactly are we predicting? What data is allowed to influence the prediction? And what happens when we act on the prediction?

What exactly are we predicting? The label is the business

The label—what you call "success" or "failure"—isn't a technical detail. It is the business decision, smuggled into a spreadsheet.

If you train a model to predict "did the customer ever miss a payment," you may reject customers who miss once but fully recover. If you predict "charge-off," you may approve customers who struggle for months, generating collections costs and customer harm, yet ultimately avoid charge-off. If you predict "fraud confirmed," you may optimize for cases that are easiest to confirm rather than those that cause the most loss. A model cannot be better than the question it is asked to answer.

Banking offers a particularly sharp version of this problem because outcomes are delayed. A loan originated today may default years from now. In the meantime, the economy changes, the borrower's life changes, and the bank changes its own policies. The label is a moving target strapped to a moving vehicle.

Data leakage: when the model cheats and you applaud

If there is a single ML mistake that deserves to be famous in banking, it's data leakage: accidentally letting the model see information that would not exist at decision time.

Leakage is how you build a model that appears brilliant and then collapses in production like a stage prop. You train a default model and include a variable like "number of missed payments in the last 60 days." Congratulations: you have built a model that detects delinquency, not a model that predicts it. It will score extremely well in testing because you gave it a cheat sheet. It will fail in real underwriting because those missed payments haven't happened yet.

Leakage can be subtle. A timestamp that encodes post-decision behavior. A "customer status" flag updated after a collections event. A fraud label influenced by the bank's own prior declines (if you decline a transaction, you may never learn whether it would have been fraud). In banking, data is full of these booby traps because operational systems are built to run the business, not to teach models clean causal stories.

The uncomfortable truth: the easiest way to improve a model is to accidentally cheat. That's why experienced teams are paranoid about how datasets are built, not just which algorithm is chosen.

Feedback loops: when prediction changes the world

Predictions are not passive in a bank. A credit model influences who gets approved; who gets approved influences who becomes a customer; who becomes a customer influences the data collected; and the next model learns from that data. The bank is a loop, not a laboratory.

This creates a weird kind of self-fulfilling pattern. If a model declines a particular population aggressively, the bank will have little data on how that population would have performed if approved. The model can become more confident over time, not necessarily because it is correct, but because it is starving itself of contradictory evidence.

Fraud systems create similar loops. If a model declines borderline transactions, it reduces fraud losses but also reduces confirmed fraud labels, because the fraud never happens. Some "wins" look like improved accuracy when they're actually improved prevention—and prevention changes the label distribution. Teams that don't understand this end up debating model quality with statistics that quietly changed their meaning.

Unsupervised learning: when you don't have labels (or don't trust them)

Not every important banking problem comes with a neat answer column. Fraud rings evolve. Money laundering patterns mutate. Operational errors appear in new forms. In these settings, waiting for "ground truth labels" is like waiting for a fire to politely file paperwork before you pick up the extinguisher.

Unsupervised learning is what you use when you want the model to find structure without being told what the structure is. Instead of learning "cat vs. not cat," it tries to discover clusters ("these transactions behave like one another") or anomalies ("this looks unlike the rest").

In fraud and financial crime, unsupervised techniques can flag transactions or accounts that are statistically unusual: sudden changes in behavior, new device-to-account connections, bursts of activity at odd hours, or networks that resemble known bad structures even when the specific actors are new. The model is not declaring "fraud." It is saying, "This is weird in a way that deserves attention."

That distinction matters. "Weird" is not illegal, and it's not always risky. The best unsupervised systems behave like good security guards: they notice what's different, they don't assume guilt, and they provide enough context for a human or downstream system to decide the appropriate response.

Unsupervised learning is also used for segmentation—grouping customers by behavior rather than by demographics. That can help with product design and customer support. But it carries a subtle hazard: behavior-based clusters can correlate with sensitive characteristics. A bank may think it's segmenting by "financial lifestyle" and inadvertently create groups that map onto protected classes. When analytics touch customer treatment, "we didn't use demographic fields" is not the same as "we didn't create demographic effects."

The two most common ML myths in banking

Myth 1: ML is about fancy algorithms. Most real ML advantage comes from unglamorous work: defining the right outcome, building a clean dataset, preventing leakage, and ensuring the model will still make sense when the world shifts. A modest algorithm on excellent data routinely beats a sophisticated algorithm on sloppy data. In a regulated industry, "explainable and stable" often outperforms "clever and fragile," even if the clever model wins a benchmark on paper.

Myth 2: ML replaces judgment. ML replaces *some* judgment with measurement, but it also creates new judgment calls: which errors matter most, what trade-offs are acceptable, when to override the model, how to monitor drift, how to respond when performance degrades, and how to communicate reasons to customers and regulators. The human role doesn't disappear; it moves up a level, from making individual decisions to designing and governing decision systems.

Generalization: the bank's real test is tomorrow, not yesterday

A model that performs well on historical data has passed a preliminary exam. Banking demands the final exam: will it work on a future that looks different?

The ML term for "working on new data" is *generalization*. The enemy is distribution shift: the idea that the world that produced your training data is not the world you're deploying into. In banking, shift isn't an edge case. It's the default. Interest-rate regimes change. Stimulus programs appear and disappear. Fraud patterns adapt to controls. Underwriting policy changes alter the applicant pool. A new digital channel attracts a different customer mix. Even the bank's own success can change behavior: customers treated well behave differently than customers treated poorly.

This is why careful dataset splitting matters so much. If you randomly shuffle years of data and split 80/20, you may end up training on the future and testing on the past, which is a time machine disguised as science. Proper evaluation respects time: train on earlier periods, test on later periods. If the model can't survive the calendar, it won't survive production.

Generalization also fails when the unit of analysis is mishandled. If the same customer appears in both training and test sets (through multiple accounts or repeated applications), the model can "recognize" rather than "learn." It's like letting a student take the exam with the answer key hidden in their pocket. In consumer credit, this can happen easily if you split at the application level instead of the customer level. In fraud, it can happen if device identifiers leak across splits. The model looks brilliant because it memorized identities, not patterns.

Features: turning raw banking data into signals

If ML is a rule-finder, features are the clues. A feature is a piece of information the model uses to make a prediction: income, utilization rate, number of recent address changes, velocity of transactions, time since account opening, ratio of deposits to withdrawals, device consistency, merchant diversity, cash-flow volatility.

Feature design is where domain knowledge shows up. A generic ML engineer might see "balance" as a number. A banking practitioner sees the rhythm behind it: payday spikes, bill-pay drops, end-of-month stress, seasonal patterns for small businesses. Two customers can have the same average balance and completely different financial lives. Features that capture volatility, trend, and timing can separate stability from fragility in ways that a single snapshot cannot.

Good features are also where ethics and compliance become concrete. Some variables are prohibited or sensitive. Others are technically permissible but operationally dangerous because they encode protected characteristics or create disparate impact. And even when a feature is legally allowed, it may be reputationally indefensible ("We denied your loan because your phone battery is usually low"). ML expands what can be measured; it does not expand what should be used.

A quick intuition for model complexity: the temptation and the trap

More complex models can capture more nuanced patterns. They can also capture more nonsense.

In lending, complexity can help discover interactions: maybe high utilization is risky only when combined with recent inquiries; maybe self-employment is fine when cash-flow stability is high; maybe a thin credit file is less concerning when deposit behavior is steady. These are plausible, valuable patterns that linear rules struggle to represent.

But complexity also increases the chance of learning spurious correlations—patterns that exist in the data by accident, or because of how the bank historically operated, rather than because of borrower behavior. Complexity can also make it harder to produce intelligible reasons for decisions, harder to validate, and harder to monitor.

Banks therefore tend to treat model complexity like leverage: useful, profitable, and requiring discipline. The goal is not to build the smartest possible model. The goal is to build a model that earns its keep under scrutiny and continues to behave when conditions change.

Where ML fits in the bank's spread machine

Machine learning doesn't change the basic economics of banking; it changes the precision of the machinery. Better prediction can widen the spread in ways customers actually feel: fewer good borrowers rejected, fewer bad loans approved, fewer legitimate transactions declined, fewer fraud losses passed on through fees and rates, faster service with fewer errors.

It also introduces a new kind of risk: model risk. A bad model can quietly scale bad decisions. In a manual process, a misguided policy might be constrained by human friction—people question it, exceptions occur, frontline staff push back. In an automated system, the same misguided logic can run cleanly and consistently, which is exactly what you want from automation *unless it's wrong*.

That's the heart of demystifying ML for banking: it is not magic, and it is not merely math. It is a way of turning historical patterns into operational decisions. The power comes from scale and subtlety. The danger comes from scale and subtlety. The bank's edge is not owning a model; it's owning the discipline to build, test, monitor, and govern models so they behave like reliable machinery rather than unpredictable creatures.

1.3 Evaluating and Interpreting AI

A model can be "accurate" and still be a financial liability. In banking, that's not a philosophical statement—it's a budget line. If the model misses the one fraud pattern that drains $20 million, or declines thousands of legitimate card transactions on a Saturday afternoon, or quietly reshapes a credit portfolio until losses spike in the next downturn, the accuracy percentage on a slide deck becomes a kind of dark comedy.

Accuracy is seductive because it's easy to say out loud. It's also strangely easy to game without intending to. Suppose only 1 out of every 1,000 transactions is fraud. A model that calls everything "not fraud" is 99.9% accurate and completely useless. Banking has many problems with that lopsided structure: rare events, high stakes, and an operational environment where the model's output changes customer behavior and therefore changes the data. A single number cannot carry all that weight.

The useful question is not "How accurate is the model?" It's: *Which mistakes does it make, how often, and what do those mistakes cost?* Models don't fail in the abstract. They fail as angry customers, regulatory findings, charge-offs, operational overload, reputational damage, and sleepless nights.

The business cost of being wrong

Every binary decision model—approve/decline, fraud/not fraud, churn risk/not churn risk—creates four possible outcomes. Two are victories. Two are headaches. And in banking, the headaches don't cost the same.

When a fraud model correctly stops fraud, that's a win you can measure: losses avoided, claims reduced, fewer compromised accounts. When it correctly lets a legitimate transaction through, that's also a win, though it's harder to celebrate because "nothing bad happened" rarely gets a parade.

The other two outcomes are where the model earns its reputation. If it blocks a legitimate customer, you create friction—sometimes minor ("please confirm this purchase"), sometimes infuriating ("your card was declined at the pharmacy"). If it lets fraud through, you create losses, operational cleanup, and the slow erosion of trust that makes customers quietly move their primary account elsewhere.

A lending model has the same four-way structure, just with different pain. Approving a borrower who repays is revenue and relationship. Rejecting a borrower who would have defaulted is a loss avoided. Rejecting a borrower who would have repaid is opportunity cost and potential fairness risk. Approving a borrower who defaults is direct credit loss plus downstream costs: collections, charge-offs, higher capital requirements, and sometimes headlines.

That trade-off deserves a visual that feels like a business meeting, not a statistics lecture.

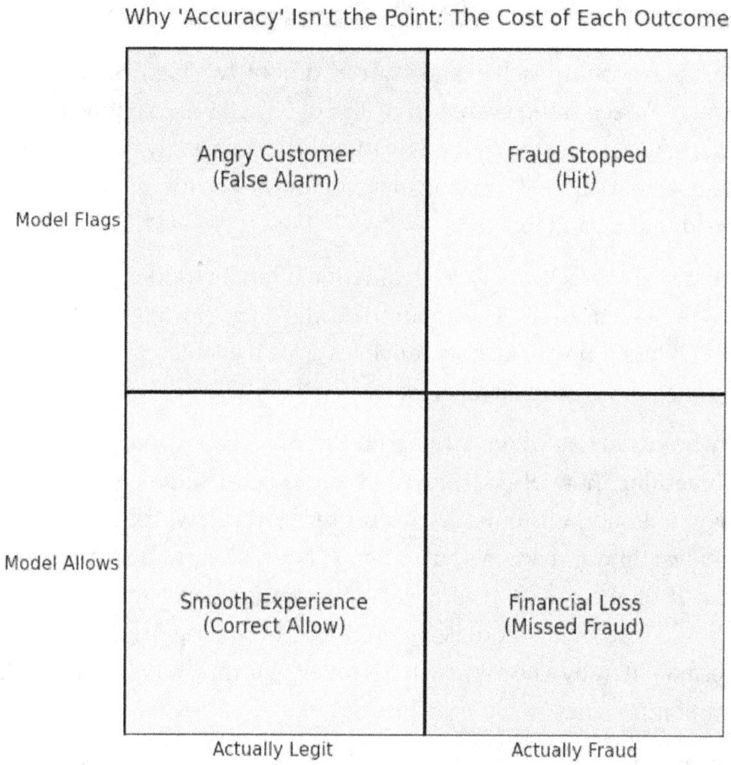

Once you see decisions this way, the conversation changes. The model isn't "good" because it has a high metric; it's good if it makes the right *trade-offs*. The bank is choosing how much customer friction it is willing to create to prevent loss, and how much loss it is willing to tolerate to avoid turning the product into a locked door.

And that choice isn't the same across products. A card network might tolerate more friction on a suspicious $8,000 transaction than on a $12 coffee. A lender might accept higher false declines for a product with thin margins and high losses, and a different balance for a relationship product where the customer lifetime value is large. Metrics only make sense in context of the operating model.

Metrics that actually behave like money

Accuracy counts each mistake as one unit of "wrong." Banking does not. A missed fraud event can cost $5 or $50,000. A wrongly declined mortgage can cost a lifetime relationship. A wrongly declined debit transaction can cost a customer's confidence at a moment when they needed the bank to work.

That's why practical model evaluation often looks like finance wearing a statistics hat. The model is judged by expected value: how much loss it prevents, how much revenue it enables, and what operational and customer costs it creates.

In fraud, teams often care about *precision* (when the model flags something, how often is it truly fraud?) because precision maps to workload and customer harm. If precision is low, the bank wastes analyst time and annoys customers. They also care about *recall* (how much of the fraud does it catch?) because recall maps to losses avoided. Tuning a model is often the art of trading precision against recall—literally choosing where to set the threshold at which the bank intervenes.

In lending, similar ideas appear with different labels. A credit model that is conservative may reduce charge-offs but also reduce originations and shrink interest income. A more permissive model may grow the book but bring losses and higher capital needs. The "best" model depends on funding costs, competitive dynamics, loss appetite, and the bank's capacity to manage delinquencies when the cycle turns. Model evaluation becomes portfolio economics.

A quietly powerful concept here is the *base rate*: how common the event is in the first place. Fraud is rare; defaults are less rare but still a minority in most portfolios. When base rates are low, even a strong model can look "wrong" a lot. That's not the model failing; it's arithmetic. If only 0.2% of transactions are fraud, a model that flags 1% of transactions and is 20% precise might still be doing valuable work—

if the dollars it saves exceed the friction it creates. Good evaluation keeps one foot in probabilities and the other in P&L.

Thresholds: where math turns into policy

Most models don't output a simple yes/no. They output a score: a probability of default, a fraud likelihood, a risk ranking. The bank then chooses a threshold: above this score, flag; below it, allow.

That threshold is not a technical afterthought. It is policy. It encodes risk appetite, customer experience philosophy, staffing levels, and sometimes regulatory tolerance. Two banks can run the same model and behave like completely different institutions because they draw the line in different places.

Thresholds also reveal a useful truth: model performance is not a single point. It's a curve. Move the threshold and you slide along a trade-off surface—catch more bad outcomes but create more false alarms, or reduce friction but accept more loss. A mature bank doesn't ask for "the best model." It asks for the model whose trade-off curve is most favorable *given the bank's constraints.*

Those constraints can be unexpectedly mundane. If an AML team can only investigate 500 alerts a day, a model that produces 5,000 alerts isn't "more sensitive." It's nonfunctional. If a call center can't handle a wave of "why was my card declined?" calls, the fraud model's threshold has a customer service budget embedded in it. Evaluation, in other words, is not only about statistical goodness; it's about operational fit.

Validation: the model has to survive an unfriendly audience

Banks treat model risk as a first-class risk because models can be wrong in two distinct ways: the model can be flawed, or the model can be used incorrectly. Both happen with depressing regularity, especially when organizations confuse "we deployed a model" with "we built a controlled decision system."

A strong model governance culture insists on *effective challenge*: an independent, informed party must be able to interrogate the model's assumptions, limitations, and performance. That sounds adversarial, but it's actually protective. It forces clarity about what the model is doing and what it is not doing.

Effective challenge also has a social function. A model built by a small team can gain a mythic status inside a large bank—"the model says no." Validation punctures that myth and replaces it with something healthier: "the model is a tool with documented limits, monitored performance, and defined overrides."

This is where documentation stops being a bureaucratic chore and becomes a survival skill. A model that cannot be explained to validators, auditors, or regulators is a model that will eventually be shut down—or worse, quietly allowed to run until it creates a loss event that makes everyone wish it had been shut down.

Explainability: not optional, not solved by a diagram

There's a temptation to treat explainability as a feature you can bolt on: add a dashboard, generate a few "top reasons," and call it transparency. Banking is not that forgiving.

In U.S. consumer lending, explainability connects directly to law. When a bank takes adverse action—declines an application, offers less favorable terms—it generally must provide specific reasons. The reasons must be meaningful to the consumer and tied to the factors actually considered. Vague explanations ("credit score too low") are not the spirit of the requirement, and in many cases are not sufficient.

This requirement changes what "acceptable" ML looks like. If a model is so complex or so opaque that the bank cannot reliably identify the principal factors driving a decision, the bank has not discovered a loophole. It has discovered a governance problem.

Even outside lending, explainability matters because banks operate in a world of dispute. Customers contest fraud claims. Businesses question credit line reductions. Regulators ask why certain groups experienced different outcomes. Executives want to know whether the model is making money for the right reasons or simply riding a temporary artifact of the data. Explainability is how the bank keeps decisioning from turning into a superstition.

Two kinds of explanations: global and local

It helps to separate explanations into two flavors.

Global explanations describe how the model behaves in general. What features matter most overall? How does risk change with utilization? Does the model treat recent delinquencies as a cliff or as a slope? Global views are useful for governance, policy alignment, and sanity checks. They answer: "What kind of machine did we build?"

Local explanations describe a single decision. Why did the model flag this transaction? Why did this applicant receive this pricing tier? Local explanations are what customers and frontline staff feel. They answer: "Why did this happen to *me*?"

Many modern explainable AI (XAI) tools—such as feature attribution methods—try to produce local explanations by estimating which inputs contributed most to a particular score. These tools can be genuinely useful. They can also be dangerously over-trusted. An explanation that looks plausible is not necessarily faithful to the model. An explanation that is faithful for one model may be unstable for another. And explanations can be manipulated, intentionally or unintentionally, by the way the model is trained and by correlations in the data.

A healthy posture is to treat XAI outputs as evidence, not verdicts. They are instruments on the dashboard, not the steering wheel.

Stress testing the explanation, not just the model

Model evaluation usually asks, "Does the prediction work?" A more mature approach also asks, "Does the explanation behave?"

If a local explanation changes wildly when an applicant's income is adjusted by $1, that's a warning sign. If two nearly identical transactions receive opposite explanations, that's a warning sign. If the model relies heavily on a feature that is operationally noisy or easily gamed, that's a warning sign. If the model's "top reasons" are things the bank cannot defend to a customer, that's a warning sign even if the model is profitable.

Banks have learned, sometimes painfully, that a profitable model can still be unacceptable if it cannot be governed. Governance is not a moral garnish; it is how the bank keeps the spread machine from being eaten by its own complexity.

Fairness, privacy, security, robustness: evaluation is bigger than a single metric

A good bank model is not only accurate. It is also safe to deploy.

Fairness matters because lending and customer treatment sit inside a legal and reputational minefield. Models trained on historical outcomes can inherit historical inequities. Even if protected-class variables are excluded, correlated variables may recreate similar effects. The bank must test for disparate impact and understand where differences come from, because "we didn't mean to" does not undo harm.

Privacy matters because models hunger for data, and banks are stewards of some of the most sensitive data people have. A model that requires invasive signals to gain a tiny lift is often a bad trade. A model that leaks sensitive information through poor controls is worse than bad—it's a crisis.

Security matters because models can be attacked. Fraudsters are professional adversaries. They probe thresholds, reverse-engineer patterns, and exploit blind spots. A fraud model that performs well against last year's data but is brittle under adaptive attack is not a fraud model; it's a training exercise.

Robustness matters because the bank's environment is not stationary. Models drift. Data pipelines change. Economic conditions shift. Vendors update software. A model that cannot be monitored and recalibrated is like a smoke detector with a sealed battery: it works until it doesn't, and you usually find out at the worst possible moment.

Frameworks like NIST's AI risk lens are useful not because they add bureaucracy, but because they force a simple discipline: map what can go wrong, measure it, and manage it as a lifecycle. Banks already do this for credit, liquidity, and operational risk. AI is just a new place where the same old law applies: unmanaged risk eventually introduces itself.

The ``black box" problem is often a ``black process" problem

People blame opacity on algorithms. Often, the deeper issue is that the *process* is opaque.

If a model's training data cannot be reconstructed, you can't prove what it learned. If feature definitions are not versioned, you can't tell whether a performance change is real or just a pipeline update. If business overrides are undocumented, you can't separate model behavior from human intervention. If monitoring is weak, you can't detect drift until it shows up as losses. A bank can buy the most interpretable model on earth and still run it as a black box by neglecting data lineage, change management, and auditability.

This is why experienced model risk teams ask boring questions with sharp edges: Where does the data come from? Who owns it? How is it transformed? What happens when a field goes missing? What is the fallback? Who can change the model? How are changes approved? Where are decisions logged? How do we reproduce a past decision? These questions sound procedural until you realize they are the difference between a controllable system and a liability that happens to output numbers.

A practical definition of ``good" in banking AI

A good model in banking is one that makes money (or saves money) *in the real operating environment* while staying governable under scrutiny. It produces the right errors at the right rate, given the bank's appetite and constraints. It remains stable enough to be trusted and flexible enough to be updated. It can be explained well enough that adverse actions are defensible, customers are treated with basic dignity, and risk committees can sleep.

That might sound like a high bar. It is. Banks don't get to run models as science projects. The model's output is a decision about someone's money. Decisions about money have a way of becoming decisions about people's lives, and people's lives have a way of showing up in courtrooms, regulatory exams, and headlines.

The good news is that this bar is also a competitive advantage. Many organizations can train a model. Fewer can evaluate it like grown-ups: with cost-sensitive metrics, disciplined thresholds, independent challenge, strong monitoring, and explanations that withstand real questions from real humans. In banking, that evaluation discipline is not a tax on innovation. It is what turns innovation into something durable.

1.4 The Rise of Generative AI in Banking

For most of modern banking history, "AI" meant something quietly mechanical: a score, a rank, a probability. A credit model takes a file and returns a number that feels like weather—*70% chance of rain, bring an umbrella*. A fraud model watches a card swipe and whispers *approve* or *decline*. Even when the math underneath is sophisticated, the output is intentionally small. Banking loves small outputs because small outputs are governable: you can set thresholds, measure error rates, and attach dollars to mistakes.

Generative AI shows up with a completely different personality. Instead of a number, it offers a paragraph. Instead of a single decision, it offers a rationale, a summary, a draft email, a call script, a policy interpretation, a plan. It doesn't just *predict*—it *produces*. That shift sounds cosmetic until you watch what happens inside a bank: whole workflows are made of words. Loans are wrapped in narratives (borrower explanations, appraisals, covenants). Fraud investigations are written up like detective reports. Compliance is largely the art of turning messy reality into tidy documentation. Customer service is, in practice, an endless river of language. A machine that can handle language at scale lands in banking like electricity—it doesn't replace one tool; it rewires the building.

Large Language Models (LLMs) are the engine behind this. At their core, an LLM is a model trained to predict the next token—a token being a chunk of text (sometimes a word, sometimes part of one). That premise sounds humble, almost silly, and it's worth sitting with. If you can predict the next token well enough across enough contexts, you aren't merely guessing words. You're learning patterns about how humans compress meaning into language: how arguments are structured, how instructions are given, how policies are phrased, how questions imply hidden needs. The model doesn't "know" in the human sense; it approximates the texture of knowledge as it appears in text.

Banking has had statistical models for decades that also predict something from something else. The difference is *what* they are trained on and *what* they are expected to output. A traditional credit model is typically trained on structured, tabular data: incomes, balances, delinquency flags, utilization ratios, time at job, and a target label like "defaulted within 12 months: yes/no." It's optimized for being right about that narrow outcome. An LLM is trained on a mountain of language and asked to be broadly competent at generating language in response to prompts. One is a specialist with a single instrument. The other is a generalist with a pen.

That difference in output changes the whole risk profile. A predictive model can be wrong, but it's wrong in a bounded way. If a model assigns a borrower a 3% probability of default and the borrower defaults, the model is "wrong" in the sense that probability is not fate; the framework accepts that some good-looking loans go bad. The failure mode is statistical. With generative AI, the failure mode is rhetorical. The model can state something false with perfect grammar and a tone that feels authoritative. In a bank, that's not a minor bug. Confidence is a feature in customer-facing language, and it becomes a hazard when it's attached to invented facts.

The industry has already given this hazard a name: *hallucination*. It's a polite term for something sharper: fabrication. A hallucinating model might cite a policy that doesn't exist, invent a fee, misread an interest rate, or produce a compliance explanation that sounds plausible but is legally off by one crucial clause. And because the output is fluent, the human reader's guard drops. Banking is full of smart professionals who are trained to mistrust ambiguous numbers; far fewer are trained to mistrust a well-written paragraph.

There's another trap that sits beside hallucination: anthropomorphism. People treat language as a sign of mind. When a system writes like an analyst, users begin to assume it reasons like an analyst. They over-delegate. They stop checking. They start asking the model to do things it can mimic but not reliably *ground*. In banking, over-delegation is a fast route to the phrase that ruins a career: "We relied on the model."

None of this means LLMs are a gimmick. It means they are a new kind of machine, and banks need to place them where their strengths are real and their failure modes are containable. A useful mental model is to separate *decisioning* from *assistance*. Decisioning is when the model's output directly triggers a financial action: approve the loan, decline the transaction, file the report, close the account, quote the price. Assistance is when the model helps a human do a job: summarize a call, extract clauses, draft a response, locate relevant policy text, translate jargon, propose next questions. In regulated environments, assistance tends to win early because it creates value without turning a language generator into an un-auditable decision-maker.

The most practical near-term uses of generative AI in banking are almost unromantic: search, summarization, and extraction. That's a compliment. Banks are buried in documents: financial statements, tax returns, pay stubs, KYC files, adverse media reports, internal policies, vendor contracts, audit workpapers, and decades of emails that contain the only copy of a crucial decision rationale. Most of that information exists as text or semi-structured blobs, not neat columns. Traditional automation struggles here because it wants clean inputs. LLMs are unusually good at taking messy text and producing a structured, human-usable output.

Consider underwriting a small business loan. The "data" isn't just revenues and balances; it's explanations: why sales dipped last quarter, what a one-time expense was, whether a customer concentration risk is temporary or structural. A predictive model can take structured features and estimate default probability, but it can't easily read a borrower's narrative and turn it into a checklist of follow-up questions. An LLM can. It can scan a banker's notes and say: *You mentioned a new supplier and a spike in COGS; do we have a revised contract? Are there minimum purchase obligations?* That isn't a decision. It's leverage: the model turns prose into underwriting discipline.

Fraud and financial crime work is even more obviously language-heavy. Suspicious Activity Reports (SARs) are narratives. Case management is narrative. The evidence is often scattered across chats, emails, and transaction notes. An LLM can help investigators by summarizing timelines, clustering similar cases, and drafting consistent write-ups. Importantly, it can do this while leaving the actual judgment—*is this suspicious?*—with the investigator. That boundary matters. The bank gets speed and consistency without pretending the model has moral or legal accountability.

Customer service is the place where generative AI feels like magic, because the product is language and the cost is people's time. A well-built LLM assistant can take a customer's messy complaint (half frustration, half facts) and produce a clean internal ticket, pull relevant account context, and draft a response in the bank's tone. It can suggest next steps: refunding a fee when policy allows, freezing a card, escalating a dispute. The result is not "robots replacing humans"; it's fewer minutes spent on typing and context switching, more minutes spent on decisions that require judgment and empathy.

The temptation, of course, is to let the model talk directly to customers with minimal oversight. That's where banking's risk reality shows up. The bank's words are not casual; they can be contractual. A sentence in a chat can be interpreted as a promise. A misstatement about a fee or a repayment schedule can trigger complaints, remediation, legal exposure, and supervisory scrutiny. A model that "sounds right" is not good enough. The bank needs correctness, consistency with policy, and traceability: *why did we say this, and what did we rely on?*

This is where an unglamorous phrase becomes the core design principle for using LLMs in banks: *grounding*. Grounding means the model's responses should be tied to authoritative sources: the bank's own policy documents, product disclosures, account data, and customer-specific facts. Instead of asking the model to answer from its general training, you retrieve the relevant documents and feed them to the model as context. The model becomes less like an improviser and more like a high-speed interpreter.

A common pattern is retrieval-augmented generation (RAG). Picture a librarian and a writer working together. The librarian (retrieval system) finds the right pages from the bank's knowledge base and policies. The writer (LLM) turns those pages into a clear answer. The trick is that the librarian must be reliable: permissions must be enforced, documents must be current, and the system must avoid leaking sensitive information across customers. In other words, the "AI problem" quickly becomes an information governance problem—which is very much a banking problem.

Even with grounding, LLMs don't behave like traditional predictive models. A credit scoring model is typically deterministic given inputs: same borrower file, same score. LLMs can be probabilistic in their outputs, and small changes in prompts can shift tone, detail, and sometimes content. That variability isn't automatically bad; it's part of why they're useful for drafting. But variability is uncomfortable in banking operations, where consistency is a form of safety. If two customers ask the same question, the bank wants the same answer, not two equally polite contradictions.

So banks wrap LLMs in controls. They constrain prompts, fix temperatures (the parameter that influences randomness), enforce templates, and post-process outputs. They add guardrails: forbidden phrases, required disclosures, escalation triggers, and citations to policy paragraphs. These guardrails are not cosmetic. They are the difference between "helpful assistant" and "uncontrolled spokesperson."

There's also a less discussed distinction between predictive AI and generative AI: what "evaluation" looks like. Predictive models are evaluated with metrics that can be plotted cleanly: ROC curves, precision/recall, false positive rates, expected loss. You can run a backtest and get a number. Generative AI evaluation is messier because the output is open-ended. What does it mean for a summary to be "correct"? You can measure factual consistency, but you also care about completeness, tone, regulatory phrasing, and whether it nudges a customer toward a safe action.

In practice, banks end up evaluating generative systems like they evaluate people: sampling, quality assurance, and supervision. They create test sets of real customer queries and internal tasks, then grade outputs against rubrics. They track categories of errors: factual errors, policy errors, privacy leaks, tone issues, over-promising, missing required disclosures. The bank's best model governance teams treat these errors the way operational risk teams treat process failures: classify, quantify, remediate, monitor.

That word *governance* matters because regulators and supervisors have not granted generative AI a special exemption. The bank is still responsible for outcomes. Many of the risks introduced by LLMs are familiar risks in new clothing: model risk (does it behave as expected?), data risk (is the data accurate and permitted?), privacy (does it leak?), cybersecurity (can it be manipulated?), and third-party risk (who runs the infrastructure?). Generative AI adds its own flavor: hallucinations, prompt injection (where a malicious input tries to override instructions), and the operational dependency on external AI providers.

Provider dependency is more than a procurement headache. If many banks rely on the same small set of model providers or cloud platforms, an outage or vulnerability can become systemic. If many institutions use similar models trained on similar data and guided by similar prompts, they can start to behave similarly under stress. Homogeneity is a quiet form of correlation risk: different banks making the same choices for the same reasons at the same time. The financial system has learned, repeatedly, that correlated behavior is where small problems become big ones.

Cyber risk also looks different when language becomes an interface. A traditional fraud model doesn't "read" an email from an attacker and decide to reveal customer information. An LLM might, if it's embedded in workflows and not properly constrained. Prompt injection is a fancy term for social engineering adapted to machines: the attacker crafts text that tries to trick the model into ignoring rules. If an LLM is allowed to call tools—query databases, send emails, update tickets—then the stakes rise quickly. It's one thing for a model to draft a bad paragraph. It's another for it to take actions.

That brings up the most important architectural choice banks make with generative AI: whether the model is *read-only* or *agentic*. A read-only system can see information and propose text, but it cannot execute. An agentic system can take steps: look things up, run calculations, initiate processes. Agentic systems are powerful, but in banking they require controls that look a lot like controls for traders: limits, approvals, logging, and separation of duties. "AI agent" is a trendy phrase; in a bank it should immediately translate into "what's the permission model, and who signs off?"

Used carefully, generative AI can do something that predictive models never could: it can make the bank feel coherent. Banks are often inconsistent because they are made of silos. One department calls it a "hardship plan," another calls it a "payment arrangement." One script says "dispute," another says "chargeback." Customers experience these inconsistencies as a lack of honesty or competence, even when the underlying issue is simply organizational sprawl. An LLM that is grounded in official language and used to generate standardized communications can reduce that sprawl. It can help ensure the bank speaks with one voice.

It can also make the bank more humane, which is not a phrase often associated with automation. The best customer interactions are clear, specific, and calm. They use plain language. They explain the "why," not just the "no." They offer next steps. Many bank communications fail here because they're assembled from legacy templates and legal caution, producing sentences that read like they were designed to survive a lawsuit rather than help a person. A generative system can draft plain-language explanations that are still compliant, especially when it is trained or prompted to respect required disclosures and avoid over-promising.

The flip side is that personalization can become manipulation if incentives are misaligned. An LLM can tailor language to what a customer seems likely to accept. That can be good—explaining a product in the way a customer understands. It can also be ethically gray—nudging a customer into a higher-fee product, or framing choices to reduce churn rather than improve outcomes. Because the system operates in language, it can influence behavior subtly. Banks will need to treat LLM-driven personalization as a conduct risk topic, not just a marketing optimization.

Internally, one of the most immediate productivity wins is in turning banking's "paperwork" into something closer to flow. Think of all the micro-tasks that fill a banker's day: reading a long email thread to figure out what was decided, scanning a policy PDF for the one paragraph that matters, rewriting a customer's request into the bank's required format, summarizing a call for the file, extracting key terms from a contract, drafting a memo for a credit committee. These tasks are individually small and collectively enormous. Generative AI is unusually good at shaving minutes off each one, which is how you get hours back without changing the bank's products at all.

There's a reason this assistive pattern shows up first in regulated environments: it fits neatly into existing model risk management frameworks. Banks already have ways to govern models that affect decisions, and those governance processes can be extended to generative AI: inventory the use cases, classify risk, validate performance, monitor drift, manage change control, document limitations, and ensure accountability. The model may be new; the need for discipline is not.

What changes is the texture of documentation. For a credit model, documentation might focus on feature selection, training data windows, stability, and performance metrics. For a generative system, documentation needs to include prompts, retrieval sources, guardrails, prohibited content, escalation logic, and human-in-the-loop requirements. A prompt can be as consequential as a line of code. A change to a policy document can change outputs instantly. Model governance becomes more like product governance: managing a living system made of model + data + instructions + tools.

One of the most important cultural shifts is accepting that an LLM is not a database. Banks have spent decades building systems where the correct answer is stored somewhere and the job is to retrieve it. LLMs don't retrieve; they *compose*. Composition is powerful but dangerous if users assume it's retrieval. A bank that rolls out generative AI needs to train staff in a new kind of skepticism: *treat the output as a draft until verified*. That training is as important as the technology.

Banks also need to decide what they want from language. Do they want speed? Consistency? Better customer experience? Lower operational cost? Fewer errors? The honest answer is "all of the above," but trade-offs appear quickly. A system optimized for speed may be more willing to answer with limited context. A system optimized for safety may be more conservative, asking clarifying questions or escalating to a human. Customers often prefer fast answers, right up until the moment the answer is wrong. The bank's job is to make that trade-off explicit rather than accidental.

The exciting part is that generative AI doesn't compete with predictive models; it complements them. Predictive models are still the best tools for turning structured signals into calibrated risk estimates. LLMs are better at turning messy language into structured inputs and turning structured outputs into human-usable explanations. A credit model might say: *probability of default increased due to high utilization and recent delinquency*. An LLM can translate that into a customer-friendly letter, and it can help an agent explain what actions could improve eligibility. It can help a banker prepare for a conversation: *Here are three questions that would clarify whether the delinquency was a one-off event*. The predictive model remains the instrument; the LLM becomes the interface.

When this pairing is done well, it makes risk management feel less like a black box and more like a process a human can participate in. That's the hidden promise: not that AI will decide everything, but that it can make decisioning systems legible—to employees, to customers, and to auditors. Legibility doesn't eliminate risk; it makes risk manageable.

The bank that thrives with generative AI won't be the one that puts a chatbot on the homepage and calls it transformation. It will be the one that treats language as infrastructure. It will build grounded systems that know what they're allowed to say, can point to the source of truth, and can gracefully admit uncertainty. It will keep humans where humans are essential: accountability, judgment under ambiguity, and moral responsibility for outcomes. And it will use machines where machines shine: reading at scale, drafting without fatigue, and turning the bank's sprawling internal knowledge into something usable in the moment a person needs it.

Chapter 2

The Engine Room: Data Platforms and Trust

If an AI model is a high-performance race car, the data platform is the engine, the fuel, and the pit crew combined. Too often, banks build gleaming models that sit in the garage because they can't be safely connected to the road. In this chapter, we move away from the hype to look at the plumbing. We will explore how to pipe data from forty-year-old mainframes to the cloud without leaks, why 'boring' governance is actually your safety net, and how to stop your AI from cheating on its tests.

2.1 Data Governance and Lineage: The Farm-to-Table Approach

Banks love the word *data* the way restaurants love the word *fresh*. It sounds wholesome. It suggests competence. It belongs on a sign in the window. But the customer doesn't actually care whether your data is "fresh." They care whether it's *true*—and regulators care whether you can *prove* it.

A useful mental model comes from food, not finance: farm-to-table. When a restaurant claims the scallops are local, it's making a promise about origin, handling, and custody. If somebody gets sick, you don't want a shrug and a guess. You want a chain of evidence: which boat, which dock, which refrigerated truck, which kitchen station, which cook, which plate. Data lineage is the same idea, minus the lemon wedge. It's the ability to take any number your bank uses—loss rate, fraud score, debt-to-income ratio, "customer lifetime value," a model's probability of default—and trace it back through every hop: where it entered, how it was transformed, and who touched it.

And the punchline is uncomfortable: without lineage, a bank isn't "data-driven." It's *rumor-driven*. The rumor just happens to be stored in Parquet.

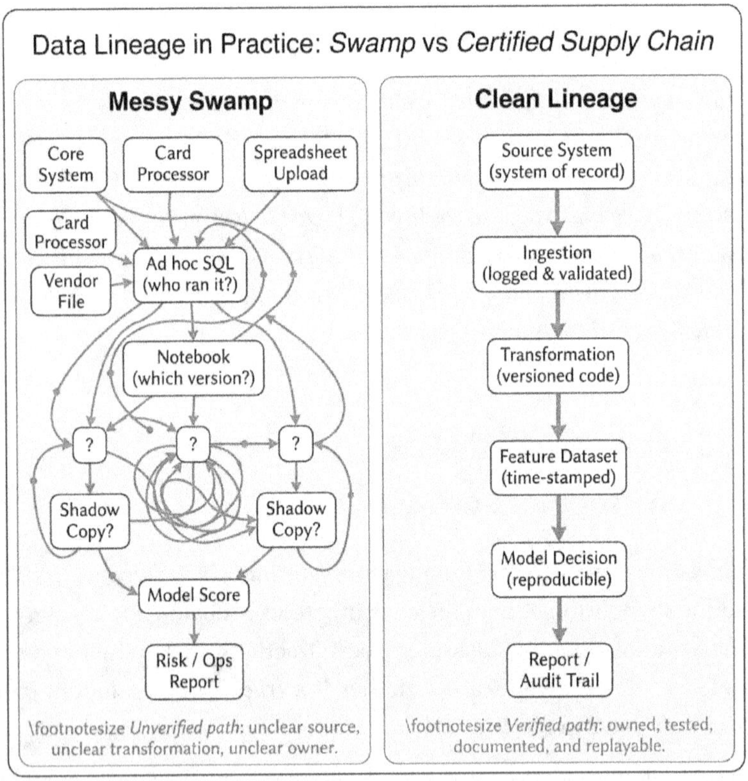

The diagram is deliberately unfair to the swamp, because the swamp is unfair to you. A swampy data platform doesn't announce itself with a warning label. It often produces numbers quickly—impressively quickly—until the day those numbers matter in a formal setting: a model validation, an exam, an audit, a disputed customer outcome, a board deck that someone will revisit after a surprise loss. Then the team is asked a simple question that feels like a trap: "Where did this value come from?" Not "generally," not "conceptually," but specifically: which system, which field, which time window, which transformation job, which version of code, which exception rules, which backfill, which manual override.

That's the moment data governance stops sounding like bureaucracy and starts sounding like insurance. Expensive insurance, yes. But the alternative is to self-insure with your reputation.

A bank's relationship with data is different from a retailer's or a media company's because bank data is routinely used as *evidence.* Evidence for why you declined a loan. Evidence for how you calculated a customer's risk grade. Evidence that your allowance for credit losses wasn't wishful thinking. Evidence that you can spot suspicious behavior without accusing innocent people at scale. The more AI you use, the more your models turn raw data into decisions that feel personal to customers and material to regulators. That shifts the question from "Is the model accurate?" to "Is the model *defensible*?"

Defensibility is not a math property; it's a *paper trail property*. Better: an *operational trail property*. Accuracy can be a lucky accident on last quarter's dataset. Defensibility requires that the bank can reproduce what happened, explain why it happened, and show that it happened the same way for everyone—especially when the outcome is unpleasant.

Consider how quickly a tiny data mystery becomes a governance problem. A credit model uses "months at current address." One applicant has "0" months even though they've lived there for years. Why? The address field was updated by customer service last week, but the model's feature pipeline reads a nightly snapshot that lagged behind. Or the address normalization service dropped apartment numbers, triggering a false "new address" event. Or a vendor feed backfilled blanks with zeros "for convenience." Each of those explanations implies a different fix, a different control, and—crucially—a different answer to the question "how many customers did this affect?" Without lineage, you can't answer that last question without a frantic archaeological dig through logs that may not exist.

Lineage turns that archaeology into a lookup.

The best way to understand governance is to stop thinking of it as meetings and start thinking of it as *ownership under pressure*. When something goes wrong, who has the authority—and the obligation—to stop the conveyor belt?

In many banks, the official answer is "the data governance committee," which is a bit like saying the fire department is "a group chat." Committees can set standards. They can't keep the kitchen clean at 2 a.m. when a batch job fails and the fraud team is screaming for scores before morning.

Operational governance means three things show up in the real world:

- A named owner for important datasets, with the power to say "no" (no silent schema changes, no mystery backfills, no "just this once" manual patches that become permanent).
- A definition of "good enough" quality that is measurable (not aspirational) and monitored like uptime.
- A lineage record that is as boring and reliable as an accounting ledger.

It sounds simple until you notice what it competes with: speed, experimentation, and the very human urge to solve today's problem with a quick extract.

That urge is how swamps are born. Not through incompetence, but through heroics.

The swamp pattern usually begins with a reasonable request: "Can we get a file of yesterday's overdrafts, split by channel?" The data team provides it, everyone celebrates, and the spreadsheet gets emailed around like contraband. Then the file becomes an input to another analysis, and that analysis becomes a slide in a deck, and the deck becomes a KPI, and the KPI becomes a target, and the target becomes a model feature. At no point did anyone say, "We are now promoting this file to a governed dataset with a controlled definition." The promotion just... happened socially.

This is why lineage is not a vanity project. It's a defense against unintentional *social truth*. In banks, the most dangerous data is not the obviously wrong data. It's the data that is *popular*.

Lineage pushes popularity back into its proper place. A dataset can be widely used and still be flagged as "unverified," "experimental," "for analysis only," or "not approved for production decisioning." Those labels are not moral judgments; they're safety markings. A bank that can't label its data can't safely scale AI, for the same reason a kitchen that can't label its ingredients can't safely scale catering.

Regulatory compliance makes the stakes sharper because regulators do not evaluate banks on vibes. They evaluate banks on controls, repeatability, and the ability to demonstrate that your internal story matches your external impact.

Model risk management—how banks control the risk of using models—treats data as part of the model, not a separate concern. A model is not just an algorithm. It is inputs, transformations, assumptions, thresholds, monitoring, and the reporting that wraps it. If the bank can't explain its inputs, it can't claim to understand its model. And if it can't reproduce the inputs used for a past decision, it can't fully reconstruct that decision when challenged.

This matters in the most awkward place: adverse outcomes. When a customer is denied credit, they often have the right to an explanation in plain language. Even if a bank uses a complex machine learning model under the hood, it still needs to produce specific reason codes—like "insufficient income relative to obligations" or "limited credit history"—that are consistent with the data used at decision time. If lineage is weak, a bank risks generating reasons that are technically plausible but not faithful to what happened. That's not just a communications problem; it's a trust problem with paperwork attached.

Lineage makes it possible to answer, confidently: "Here is the data we used, here is when we pulled it, here is how we transformed it, and here is how those values contributed to the decision."

A practical definition helps: data lineage is a map of *data provenance* (where it came from) and *data transformations* (what happened to it) from source to consumption. "Consumption" includes dashboards, regulatory reports, and AI models—anything that turns data into action.

Banks often underestimate how many transformations occur before a model sees a number. A single "utilization ratio" might involve: raw tradeline data from a bureau, parsing, deduplication, selection of "latest" records, handling of missing limits, conversion of currencies, winsorization (clipping extreme values), aggregation, and time alignment to the application date. That's before any fancy feature engineering begins.

Each transformation step is a chance to introduce errors, bias, or drift. Not necessarily because someone is careless, but because transformation steps encode policy: how to treat missing data, how to resolve conflicts, how to represent customers with thin files, how to interpret joint accounts. Those policies can have disparate effects across customer groups. If a bank can't see the policies embedded in its feature pipelines, it can't audit for fairness in a meaningful way. It can only argue about outcomes after the fact, when the argument is expensive.

Lineage is what turns "we think" into "we can show."

The farm-to-table analogy has another useful edge: it highlights the concept of *chain of custody*. In food, custody is physical: temperature logs, sealed containers, sanitation checks. In data, custody is logical: access controls, change logs, approvals, and evidence that the dataset used for production is the dataset that was tested and validated.

The ugly failure mode is a "definition fork." One team calculates "delinquency" as 30+ days past due; another uses 60+; a third excludes accounts in hardship programs; a fourth includes them but flags them. All four definitions may be defensible in isolation. The problem is when they collide in an executive dashboard or—worse—in a model training pipeline, where subtle mismatches create phantom predictive power. A model looks brilliant because a feature definition accidentally smuggled in post-event information, or because the target variable is mislabeled in a way that aligns with operational processing rather than customer behavior.

Lineage doesn't prevent forks by itself, but it makes forks visible. And visibility is the prerequisite to governance. You can't standardize what you can't see.

Lineage also changes the economics of debugging. Without it, every investigation is bespoke. A model performance dip becomes an all-hands meeting and an exchange of Slack messages that read like a detective novel: "Did we change the bureau vendor?" "Was there a system outage?" "Is the batch job late?" "Did someone patch the transformation?" "Are we still using the same population?" Each question triggers a manual query, often by the one person who knows where the bodies are buried.

With lineage, investigations become closer to incident response in modern engineering: you can trace a bad value upstream, identify the first transformation where it diverged from expectation, and isolate the blast radius. That is not just faster; it's calmer. The organizational benefit is not speed—it's less panic, and fewer "emergency" changes that create tomorrow's swamp.

Calm is underrated in financial risk.

What does "good" governance look like in a bank that actually builds and deploys AI, not just talks about it? It looks like a small set of artifacts that are always kept current because people rely on them to do their jobs. The key is to treat governance outputs as *products*, not *documentation*. Products have users. Documentation has victims.

A compact way to describe those artifacts is below:

Governance artifact	*What it really is*	*Who depends on it*
Data inventory (catalog)	A searchable "menu" of datasets with definitions, owners, and allowed uses	Analysts, modelers, auditors, exam teams
Lineage graph	A wiring diagram showing how values flow from source to transformation to consumption	Model validators, data engineers, incident responders
Data quality rules	Automated checks that turn "this looks off" into measurable thresholds and alerts	Operations, risk, engineering on-call
Business definitions	One agreed vocabulary for core concepts (e.g., delinquency, income, exposure)	Finance, risk, compliance, product

Notice what's missing: the ceremonial parts. Not because they're useless, but because ceremonies only work when the underlying artifacts are alive.

A common misconception is that lineage is something you "add" after the platform is built, like sprinkling parsley on a finished dish. In practice, lineage is easiest when it is a side effect of how work gets done. If jobs are orchestrated, transformations are version-controlled, and datasets are registered when they're created, lineage can be generated automatically. If work happens in one-off scripts on laptops and datasets are shared by email, the lineage project becomes a manual data-entry nightmare. The organization ends up hiring a small army to document a process that is still changing underneath them.

So the deeper goal is behavioral: make the governed path the easy path.

That usually means building paved roads:

- Standard ingestion patterns that automatically log what arrived, when, and from where.
- Standard transformation frameworks that record code versions, inputs, outputs, and runtime metadata.
- Standard publication steps that register datasets and attach owners, definitions, and allowed uses.

When paved roads exist, teams use them not because they're forced to, but because the roads are faster than bushwhacking.

There's also a human tension worth naming: governance is often seen as a tax on creativity. Data scientists want to try ten feature ideas before lunch. Engineers want to ship. Product teams want answers yesterday. Governance feels like someone asking for a receipt while your hair is on fire.

The trick is to separate two modes and label them honestly:

Exploration mode is allowed to be messy—as long as it is clearly marked as non-production and non-decisioning. Think of it as the test kitchen. You can experiment, but you don't serve it to customers.

Decision mode must be governed, traceable, and repeatable. Think of it as the dining room on a Saturday night. You don't invent a new refrigeration policy mid-service.

Banks get into trouble when they pretend their test kitchen is a dining room because the food *usually* tastes fine. Lineage is one of the few mechanisms that can enforce the distinction without smothering experimentation. If the platform can tag datasets as "exploratory" and prevent them from quietly crossing into production pipelines, governance becomes a guardrail instead of a speed bump.

Lineage is also a surprisingly good antidote to internal politics. When a number is disputed—say, two teams have different counts of "active customers"—the argument often becomes social: whose dashboard is trusted, whose executive sponsor is louder, whose SQL looks more sophisticated. Lineage changes the argument into engineering: which definition is used, which tables are sourced, which filters are applied, which time zone is used, which late-arriving records are included, which exceptions are carved out. People can still disagree, but at least they disagree about *something concrete.*

Concrete disagreements are progress. Vague disagreements are how banks end up with parallel realities.

There is one more reason lineage matters in AI specifically: models are exquisitely sensitive to silent change. Traditional reports can survive some instability because humans interpret them. A model does not interpret; it reacts.

Change the meaning of a single categorical value upstream—merge two merchant categories, rename a channel code, modify how chargebacks are recorded—and a fraud model may misread the world. The model's math hasn't "broken," but the bank has changed the language the model uses to think. If no one can see that language change, performance deterioration looks mysterious, and the fix becomes guesswork.

Lineage supports a more grown-up relationship with change. It enables *impact analysis*: if we change this field, which models and reports are downstream? Who needs to test? Who needs to sign off? Which monitoring thresholds should be tightened during the transition? Without lineage, banks often learn downstream dependencies the hard way: by breaking them.

None of this requires perfection. Banks don't need to trace every byte. They need to trace the data that matters—material risk measures, decisioning inputs, regulatory reporting figures, and the features used in high-impact models. A risk-based approach keeps the work sane.

What "risk-based" means in practice is that a bank should be able to point to a dataset and say:

- Whether it is used for customer decisions, financial reporting, or capital/risk measurement.
- How sensitive it is (does it contain customer information, trade secrets, or security-relevant signals?).
- How frequently it changes, and how likely it is to break downstream consumers.
- What happens if it's wrong: inconvenience, financial loss, unfair outcomes, regulatory exposure, or all of the above.

Then governance effort matches consequences. The scallops get more paperwork than the parsley.

A final, practical test separates real lineage from decorative lineage: replayability. If you take the exact inputs and the exact code versions that existed on a given day, can you reproduce the dataset the model saw? Can you reconstruct the score? Can you show the reason codes?

Can you demonstrate what would have happened if an upstream system had been delayed by two hours?

Replayability is where governance earns its keep because it forces precision. It turns lineage from "we drew a diagram" into "we can rerun the factory." That capability is priceless when a bank needs to answer a regulator quickly, respond to a customer dispute, or investigate why a model behaved differently during an incident.

Farm-to-table is ultimately about trust. Not trust as a brand feeling, but trust as an operational fact. You don't get it by saying "we take data seriously." You get it by being able to point—calmly, precisely, and repeatedly—to where each data point came from, what happened to it, and why it deserves to be on the plate.

2.2 Privacy, Consent, and Security: Keeping Secrets Secret

Banks are in the secrets business. Not as a marketing slogan—as a mechanical fact. Every day, customers hand over the details that make them legible to a balance sheet: income, rent, medical debt hints hidden in merchant names, the awkward timing of a payday loan, the quiet panic of overdrafts in the last week of the month. A bank's job is to turn those secrets into useful action without turning them into gossip.

AI complicates that job because it has a talent humans rarely do: it can absorb *everything* and still look calm. A model doesn't blush when it sees a Social Security number. It doesn't feel inappropriate. It just treats it as a token that might help predict something—and if you give it enough training time and enough parameters, it may *remember* it.

That word—*remember*—is doing new work in banking. In the old world, the nightmare was someone stealing a database. In the new world, the nightmare includes someone extracting sensitive information from a model, or having a model inadvertently leak customer details through logs, prompts, or "helpful" explanations. The perimeter is no longer just the warehouse. It's the warehouse *plus everything that learned from it.*

Privacy in banking tends to be introduced with solemnity and acronyms. Useful acronyms, but still acronyms. It's easy to treat privacy as a compliance box and security as an IT box. That mental split breaks down the minute you deploy machine learning at scale, because models create a new kind of asset: a decision engine built out of customer data. Security has to protect it. Privacy has to constrain it. And consent has to justify it—or at least keep it from becoming a trust crater.

A good guiding principle is disarmingly simple: *don't feed the model what you don't want to protect.* Data minimization means collecting and using only what is necessary for a specific purpose. It sounds like etiquette. In practice, it's a hard-nosed engineering strategy. Every extra field you pipe into training is another thing that can be misused, mishandled, misinterpreted, or memorized. The cheapest breach is the data you never stored. The easiest incident response is the one you don't have to run.

Banks often do the opposite by accident. Data platforms make it easy to say, "Just land everything in the lake, we'll figure it out later." That's how you get an AI pipeline that quietly ingests raw PDFs, call-center notes, document images, and "miscellaneous" fields that are basically free-form human confession. Then the modelers—under pressure, always under pressure—discover a feature that seems to improve accuracy. It's tempting. It's also how secrets escape: not with malice, but with curiosity and a leaderboard.

Data minimization becomes real when it is enforced at the seam where raw data meets model-ready data. Think of it as a privacy filter: a deliberate choke point where the bank decides what is permitted to cross into training and scoring environments. The filter's job is not to make data pretty. Its job is to make data *appropriate.*

Appropriate data has a few properties that sound boring until you're in front of an examiner:

- It is tied to a declared use case. "For credit underwriting" is not the same as "for marketing," and neither is the same as "for fraud."
- It is stripped of direct identifiers wherever possible. Names, full account numbers, government IDs, precise addresses—these are rarely necessary for prediction once the bank has already linked records internally.
- It is processed to remove "data shrapnel": free-text fields and attachments that can contain anything from medical details to passwords customers unwisely paste into chat windows.

Notice what's not on that list: "just hash it." Hashing (turning a value into a one-way fingerprint) is useful, but it's not a magic cloak. A hashed Social Security number can still act like a unique identifier. A model can still latch onto it. And if the hashed value is stable, it becomes a long-lived tracking tag—exactly what privacy teams are trying to avoid.

There is also the problem of *proxy data.* Even if you remove obvious personally identifiable information (PII), models can infer sensitive facts from patterns. Zip code is the classic example: it's not "race," but it can correlate strongly with race because housing patterns are not random. Merchant categories, device types, shopping times, and even the cadence of bill payments can become proxies for income volatility, health status, or other deeply personal attributes.

This is where privacy stops being a redaction exercise and becomes a risk-management exercise. You can't simply delete every field that correlates with something sensitive, because correlation is everywhere. But you can do two things that are both practical and defensible:

- Make the purpose explicit and test for harmful impact. If a feature is a strong proxy and creates disparate outcomes, it becomes a governance issue, not a clever trick.
- Prefer aggregated, behavior-level signals over raw, identity-level detail. "Number of late payments in the last 12 months" is usually safer than storing a list of exact late-payment dates with source references that point back to specific merchants or medical providers.

Banks sometimes shy away from this conversation because it sounds philosophical. It isn't. It's about whether the model is learning financial reliability or learning social geography.

Consent is the most misunderstood ingredient, partly because the word makes people picture a checkbox. In reality, consent is a relationship between a promise and a use. Customers may be comfortable with their data being used to prevent fraud—most people like not being robbed. They may be far less comfortable with the same data being used to predict what product they're "likely" to accept, or to decide how aggressively to market credit.

The problem is that AI systems are naturally promiscuous: the same dataset can be repurposed endlessly, and the marginal cost of doing so feels close to zero. A bank's data platform will happily feed a fraud feature store into a marketing model if no one stops it. That's not just a compliance risk. It is how trust erodes quietly, customer by customer, until a headline arrives and everyone suddenly remembers that "creepy" is a business risk.

The operational challenge is to encode consent and permitted uses as *controls*, not as PDFs. That means tagging data with allowed purposes, enforcing those tags through access controls and pipeline checks, and building audit trails that can answer uncomfortable questions later: Who accessed the data? For what job? Under what approval? Did it leave the environment? Did it end up in a model artifact?

A bank that can answer those questions quickly looks competent. A bank that can't looks like it is hoping nobody asks.

Security, meanwhile, has to treat AI pipelines as first-class critical infrastructure. Many organizations secure their core systems tightly and then treat analytics as a playground. That split is survivable when analytics produces internal reports. It becomes dangerous when analytics produces automated decisions or customer-facing interactions.

A modern AI pipeline has a lot of attack surfaces:

- Data ingestion endpoints (SFTP drops, vendor APIs, event streams).
- Storage layers (raw zones, curated zones, feature stores).
- Training environments (clusters with powerful compute and broad data access).
- Model artifacts (weights, embeddings, vector indexes) that are portable and easy to copy.
- Serving endpoints (real-time scoring APIs, batch scoring jobs).
- Logs (which, with the best intentions, can become a scrapbook of sensitive inputs).

An attacker doesn't need to "hack the bank" in a cinematic way. They can slip in through a neglected service account, an overly permissive bucket, or a third-party integration that was approved fast and reviewed slowly. The industry has learned the hard way that cloud doesn't remove responsibility; it redistributes it. The provider secures the building. The bank still has to lock the door to the vault.

One of the least glamorous and most effective privacy-and-security controls is a data inventory. It's hard to protect what you can't name. Regulators have increasingly emphasized that financial institutions should know what customer information they have, where it lives, and how it moves. That sounds like an administrative task until you remember that AI training often involves copying data into new places: scratch spaces, temporary tables, cached feature sets, snapshots for reproducibility. Every copy is another chance for oversharing and another location that must be secured, monitored, and eventually deleted.

Deletion matters more than people expect. "We'll keep it forever in case it's useful" is a natural instinct in data science, but it's an expensive instinct in security. Long retention increases the blast radius of any breach. It also increases the likelihood that data will be used in ways that were never intended. Minimization is not only about what you collect—it's about how long you keep it and how many clones you create.

A bank that does disciplined retention isn't being precious. It is reducing future incident costs in a way that can be measured.

The FTC's Safeguards Rule, which implements standards for protecting customer information under the Gramm-Leach-Bliley Act (GLBA), captures the mood of modern expectations: written information security programs, risk-based safeguards, oversight of service providers, and controls that match the sensitivity of the data. The details can feel legalistic, but the spirit is straightforward: if you handle customer information, you need a security program that is real, documented, and operational—not a binder that gathers dust until the week before an exam.

What changes with AI is that "customer information" isn't only in the obvious places. If a model can be probed to reveal sensitive training examples—or if logs capture raw inputs that include PII—then the security program has to expand its mental model. Protecting the warehouse but ignoring the model is like installing a state-of-the-art alarm system and leaving your keys taped under the welcome mat.

It's worth being explicit about the new failure modes AI introduces, because they're counterintuitive:

Memorization. Large models and even some smaller ones can overfit in ways that effectively store rare or unique records. If you train on small populations—say, high-net-worth customers, niche fraud patterns, or rare dispute categories—the uniqueness of the data makes memorization more likely and more damaging.

Inversion and extraction. Attackers may try to reconstruct sensitive attributes by querying a model strategically, especially if the model returns rich outputs (probabilities, confidence scores, embeddings) rather than simple decisions.

Accidental disclosure through explainability. Explainability tools are meant to make models more transparent. But a careless explanation can expose sensitive inputs ("declined because SSN matched prior fraud"), or reveal too much about the bank's internal detection logic, creating a playbook for criminals.

Logging as leakage. Engineers log things to troubleshoot. Data scientists log things to debug feature pipelines. Product teams log things to improve experiences. If raw inputs are logged, secrets have a habit of traveling to places with weaker controls.

None of these risks require panic. They require design discipline: limit what the model sees, limit what the model says, and limit what the system remembers about each decision beyond what is necessary for audit and service.

If minimization is the diet, encryption is the safe packaging. Banks already know to encrypt data at rest and in transit. The AI twist is that more systems touch the data, and more "temporary" movements happen. Training jobs copy datasets onto compute nodes. Feature pipelines stage intermediate outputs. Model deployment often involves containers, model registries, and artifact stores. Each of these is an opportunity for someone to forget to turn on encryption or to rely on defaults that are not as default as they think.

Access control is just as important as encryption, and it's where banks most often stumble—not because they don't have controls, but because they have *too many ways around them.* AI teams move fast, and they often need broad access to experiment. Broad access, in turn, tends to create permanent exceptions: the service account that "temporarily" got read access to a raw bucket; the contractor role that "needed it for a week"; the shared notebook environment where credentials are stored for convenience.

A security program that works with AI treats identities as carefully as it treats data. Principle of least privilege (give each user/system the minimum access needed) sounds like a slogan until you discover that a single overpowered account can quietly export an entire training corpus.

Third parties deserve special skepticism, not because vendors are untrustworthy, but because integrations multiply complexity. Banks rely on cloud providers, fraud platforms, KYC utilities, marketing analytics tools, call-center systems, and data brokers. Each relationship creates data flows. Each flow creates an obligation to manage risk across a lifecycle: selecting vendors carefully, negotiating contracts that specify security expectations, monitoring performance, and planning termination so data doesn't linger after the relationship ends.

AI increases the stakes because vendors may not just store data; they may train models on it, or provide models that are trained on pooled industry data. That can be valuable. It can also be confusing. Questions that matter suddenly sound technical but are actually governance questions:

- Is the vendor allowed to use your data to improve their model for other clients?

- Do you get transparency into how the model was trained and validated?
- How are outputs logged and stored, and who can access them?
- What happens to your data and derived artifacts when the contract ends?

If those questions aren't answered early, they get answered later—during an incident, when "later" is the worst time.

A subtle privacy risk shows up in the most well-intentioned place: customer experience. Banks increasingly want AI-powered assistance—chatbots, agent copilots, automated email drafting, "next best action" recommendations. These systems can be helpful, but they are also excellent at over-sharing unless carefully constrained. If the assistant can see raw customer data, it might include it in a response. If it can see internal case notes, it might repeat them. If it can see model outputs, it might expose them in ways that confuse customers or create new attack vectors.

A safe pattern is to treat customer-facing AI like a diplomat: it only speaks from prepared notes. That means:

- Restricting what data the system can retrieve (purpose-limited retrieval).
- Masking or redacting sensitive fields before they ever reach the model.
- Structuring outputs so the system cannot "freestyle" with private details.

The goal is not to make the AI less capable. It's to make it less improvisational with secrets.

Breach response is where governance, security, and privacy all meet in a room together—usually without enough sleep. Modern rules increasingly expect timely notification for certain security events, and banking organizations have to be ready to determine what was accessed, whether it was encrypted, and how many customers were affected. AI systems complicate that accounting because sensitive information might not only be in a database; it might be in an exported training set, a cached feature snapshot, or a model artifact replicated across environments.

This is another reason data lineage matters even when the topic is privacy. When an incident happens, the bank needs to know where sensitive data traveled. Not "generally," but specifically. Which pipelines had access? Which service accounts ran them? Which destinations were written? Which third parties received feeds? Without that trace, banks are forced into worst-case assumptions—expensive, reputation-damaging assumptions.

Privacy and security are often framed as constraints that slow innovation. A better framing is that they are constraints that keep innovation *in bounds*. AI that moves fast and breaks things is a cute slogan until "things" includes customer trust, fair access to credit, and the bank's ability to operate without supervisory pain.

The highest-performing AI organizations inside banks tend to develop a distinctive habit: they treat privacy and security as *engineering requirements*, not as *review steps*. That sounds like semantics, but it changes behavior. Review steps happen at the end, when changes are expensive and teams are emotionally invested. Engineering requirements shape the design up front: what data is allowed in training, how it is masked, how access is granted, how logs are sanitized, how artifacts are stored, and how the system can be audited.

That approach also produces something rare in large institutions: speed that doesn't feel reckless. When teams know the rules and the platform enforces them, they stop negotiating each project from scratch. They build within guardrails, and the guardrails are sturdy enough that nobody has to pretend they're optional.

A bank doesn't get credit for *having* secrets. It gets credit for *keeping* them. In an AI-powered bank, keeping secrets secret is no longer a side function. It's a core competency—one that determines whether machine learning becomes a competitive advantage or an unusually expensive way to create new kinds of breaches.

2.3 Integration with Core Systems: The Clash of Generations

A bank can build a gorgeous machine learning model in a modern stack—Python notebooks, feature stores, elastic cloud compute, the whole glossy showroom—and still lose to a 40-year-old mainframe at the exact moment it matters. Not because the mainframe is "better," but because it is *closer to the money*. The core system sits where balances are real, transactions are final, and timeouts are not theoretical. Your model is a guest in that world, and guests don't get to rearrange the furniture.

This is the clash of generations: the fast, expressive, experiment-happy culture of data science meeting the slow, conservative, uptime-obsessed culture of core banking. One side ships code like a news cycle. The other side measures change in maintenance windows and scars. Both are right. The trick is getting them to talk without starting a fire.

The core system—often a COBOL mainframe, sometimes a thick Java stack with decades of sediment—was built for a specific kind of excellence: correctness at scale. It posts transactions in a strict order. It guards the ledger the way a museum guards a famous painting: with boring rituals that work. If it feels stubborn, it's because it has spent its life being blamed for things it didn't do.

Machine learning systems, by contrast, are built for a different kind of excellence: adaptability. If the fraud landscape shifts, you retrain. If customer behavior changes, you recalibrate. If a feature breaks, you patch and redeploy. The model is not a monument; it's a living organism. Put that organism in front of a core system that demands predictability, and the cultural mismatch becomes technical risk.

The integration question is not "Can we deploy AI?" It's "Can we deploy AI *without increasing the probability of a bad day?*"

The easiest way to misunderstand integration is to frame it as plumbing. It isn't. Plumbing implies that water is water. In banks, data is political, time-sensitive, and sometimes wrong. Worse, it's wrong in ways that only show up when you compare systems that disagree about the present.

Consider a simple statement like "available balance." On the customer's screen, it changes in real time. In the core, it may involve posted transactions, pending authorizations, holds, and cut-off times. In a data warehouse, it might be yesterday's snapshot. In a fraud system, it might be a streaming approximation. If an AI model is scoring a transaction and needs "available balance," which version does it mean? If the model's training used a warehouse snapshot and production uses a real-time core query, the model will behave like it learned one language and is now being spoken to in another.

Integration isn't just about moving data. It's about aligning realities.

Banks usually face this problem in its most dramatic form in fraud. Fraud decisions live in a brutal time budget: milliseconds to decide whether to approve, decline, or step-up authenticate a transaction before the payment network gives up waiting. In that window, your model can't go on a leisurely tour of the enterprise. It can't query five systems, join three tables, and wait for a "helpful" API that sometimes takes 900 milliseconds because it's having a day.

If your fraud model needs to call the core to fetch features, it might work in testing and fail in production during peak load—exactly when fraud attempts spike. That failure mode is nasty: the integration becomes the bottleneck, and the bank is forced to choose between letting transactions time out (customer pain) or bypassing the model (fraud loss). A bank that can't degrade gracefully will eventually degrade catastrophically.

So integration design starts with a blunt question: *How much latency can we afford, and what happens when we don't get it?*

There are two broad patterns for integrating models with core systems: batch and real-time. Banks often treat this as a technology preference. It's really a risk preference.

Batch integration means the model runs on a schedule—nightly, hourly, whatever—and produces outputs that downstream systems consume later. Batch is slower, but it is operationally forgiving. If a job fails, you rerun it. If an upstream feed is delayed, you shift the window. Batch is how many credit risk and marketing models have lived for years: score the portfolio overnight, update segments, send files to the core or to downstream teams.

Real-time integration means the model is called during a live transaction or interaction. It's thrilling when it works. It's unforgiving when it doesn't. Real-time requires low-latency feature availability, high system resilience, careful fallbacks, and monitoring that notices trouble before customers do.

Most banks end up with both, and the tension is that the same model can't always live comfortably in both worlds. A feature that is fine in batch ("count of disputes in the last 90 days") becomes tricky in real-time if the disputes system updates late. Conversely, a real-time feature ("device fingerprint risk score") may not be stored historically in a way that supports training.

This is how integration quietly shapes what models are even possible.

A practical way to think about integration is in three layers:

- *The ledger layer* (core systems): the authoritative truth of balances, postings, and customer records.

- *The decision layer* (models and rules): the logic that turns context into action.
- *The interaction layer* (channels): mobile app, branch, call center, payment rails—where decisions must arrive on time and in the right format.

When banks struggle, it's often because they try to collapse layers. They want the model to reach directly into the ledger layer for features and then push the decision back into a channel, all in one synchronous chain. That chain becomes fragile: every additional dependency is another place to time out, another place to have a mismatch, another place to have an incident.

The more resilient approach is to decouple layers so failures are contained. Decoupling sounds abstract until you live through your first incident and realize the mainframe was never supposed to be in your model's critical path.

This is where an unglamorous but powerful pattern earns its keep: change data capture (CDC). CDC is a method for capturing changes from a database—often by reading the database's transaction log—so downstream systems can receive updates quickly without hammering the core with queries. Instead of polling ("Are we there yet?" every minute), CDC listens to what actually changed and publishes it as a stream.

CDC matters because it lets modern systems stay current without asking the core to behave like a modern system. The core keeps doing what it does best: processing transactions reliably. Downstream, a streaming layer receives near-real-time updates—new transactions, balance changes, status updates—and builds model-ready views. The model reads those views with low latency, and the core is spared a flood of feature queries.

Banks like CDC for another reason that is more subtle: it creates a *record of change*. A stream of events is naturally auditable. When a model made a decision, you can reconstruct what the world looked like from the event history—if you store it properly. That ties integration back to trust: if you can't replay the past, you can't fully explain it.

Of course, CDC is not magic. It's a commitment. You have to manage schemas, handle late or out-of-order events, and ensure that downstream consumers don't silently drift into incompatible interpretations. But compared to the alternatives, CDC is often the least bad option.

The alternative many teams reach for first is the "dual write": when a transaction happens, write it to the core *and* write it to the analytics or feature store at the same time. It's appealing because it feels direct. It's also a trap, because "at the same time" is a romantic notion, not a distributed-systems guarantee.

Dual writes break in ordinary ways: the first write succeeds and the second fails; the second succeeds and the first fails; both succeed but in different orders; retries create duplicates; a network hiccup turns a clean event into a ghost story. You can try to make it robust with distributed transactions, but heterogeneous systems—mainframes, modern databases, cloud queues—rarely play nicely in a single transactional boundary. You end up building a custom reliability mechanism that becomes its own critical system, maintained by people who would rather be doing literally anything else.

CDC, when feasible, is often safer because it treats the core as the source of truth and derives downstream state from the core's own committed history. It's not perfect, but it's aligned with how banks already think: the ledger is the ledger.

Even with CDC, integration hits a second hard problem: *identity*. Modern ML stacks like to think in terms of "entities"—customer, account, device, merchant. Core systems may have multiple IDs for the same human depending on product, channel, or merger history. A person can be a customer in one system, an applicant in another, and an "individual party" in a third. If your model's feature store assumes one clean ID and the bank's reality is three messy ones, you end up with Frankenstein profiles that combine the wrong attributes or split a person into multiple partial ghosts.

Identity resolution can be done well, but it is not a weekend task. It requires rules, governance, and a willingness to treat reference data—customer identifiers, product hierarchies, merchant mappings—as first-class assets. If those assets are unstable, models become unstable, and integration becomes a treadmill of "Why did this score change?" questions.

The most expensive AI bug is often not a bug in the model. It's a bug in *who the model thinks the customer is.*

Cloud adds another layer of temptation and another layer of responsibility. It's tempting to believe that moving the decision layer to cloud automatically makes the bank more modern, more secure, more resilient. The cloud does provide powerful tools. But it doesn't absolve the bank of governance. Cloud risk management is a shared responsibility: providers secure infrastructure, while banks remain responsible for how they configure services, manage identities, monitor access, and ensure resilience.

Integration is where shared responsibility becomes painfully concrete. The model runs in cloud. The core lives on-prem or in a hosted mainframe environment. Data moves between them. Credentials bridge them. Monitoring has to cover both. When an incident happens, it rarely respects those boundaries.

The safest stance is to assume that anything crossing the boundary will eventually be misconfigured unless your processes make misconfiguration hard. That means strong controls on network paths, encryption, key management, service identities, and vendor oversight—not as a ceremonial checklist, but as a living system that's continuously tested.

Most banks also have an emotional integration problem: fear of "big bang" modernization. They've seen transformations that promised to replace the core and instead replaced executive teams. So they stick with what works, and integration becomes a patchwork: point-to-point connections, brittle file transfers, and "temporary" middleware that turns permanent.

The antidote is incremental modernization with a clear boundary strategy. A useful pattern is often called the Strangler Fig: you gradually route parts of functionality away from the legacy system into new services, while the legacy core continues running. The "fig" metaphor is less important than the operational benefit: you avoid betting the bank on a single cutover date. You replace pieces, verify behavior, and expand the new boundary over time.

AI fits naturally into this pattern because models are usually *additive* at first. They advise, score, recommend, and flag. They don't have to immediately become the system of record. A bank can start by inserting models as decision services around the edges—fraud scoring, credit pre-qualification, collections prioritization—while keeping final posting and authoritative balances in the core.

The goal is not to avoid change. It's to make change survivable.

One of the most practical integration design choices is deciding where features live. Features are the inputs a model uses—things like "number of NSF fees in last 30 days" or "time since last password reset." If features are computed on the fly by querying source systems, your model inherits the fragility of those systems. If features are precomputed in a feature store, your model inherits the responsibility of keeping that store current, correct, and consistent.

Precomputing features is often the right answer for real-time use cases because it allows low-latency reads. But it creates new governance demands:

- Features must be time-aligned so the model doesn't accidentally use information that wasn't available at decision time.

- Feature definitions must be versioned, or model performance becomes a mystery.
- Backfills must be controlled, or yesterday's "truth" changes under your feet.

Integration is the place where data governance stops being abstract. It becomes the difference between a feature store that is a reliable instrument panel and one that is a haunted house.

Then there's the question everyone eventually asks: *Where does the decision get enforced?* A model score by itself doesn't stop fraud. It doesn't approve a loan. It doesn't reverse a fee. It becomes useful only when it is translated into an action in an operational system—often the same core or channel systems that were hard to integrate with in the first place.

This is why "just deploy the model as an API" is not a complete integration strategy. The API can return a score, but something has to:

- interpret it in the context of policy,
- apply thresholds that may differ by product or channel,
- handle exceptions and overrides,
- log what happened for audit and customer support,
- and fail safely when the model is unavailable.

A mature bank treats this as decisioning architecture, not model deployment. Often that means a decision engine that combines model outputs with rules and policy constraints, and then sends a compact, deterministic instruction to the channel: approve, decline, step-up, refer to manual review. The model informs the decision; the decision system owns the action.

This separation protects the bank in two ways. First, it reduces the temptation to let raw model outputs leak into customer-facing surfaces. Second, it creates a stable interface for the core and channels.

The model can evolve behind the interface without forcing every downstream system to change each time a data scientist retrains.

The best integration projects in banks share a slightly countercultural mindset: they optimize for *boring*. Boring means predictable latency. Boring means robust fallbacks. Boring means versioned interfaces. Boring means that an engineer can go on vacation without the system turning into folklore.

Boring also means observability—knowing what is happening as it happens. If a real-time model endpoint starts timing out, you need to see it within minutes, not after the daily report. If a CDC stream falls behind, you need alerts before the feature store becomes stale. If a schema changes upstream, you need detection before the model starts scoring everyone as a fraudster.

Observability is not a nice-to-have in integration. It is how you prevent a small mismatch from becoming a bank-wide incident.

Integration is where many AI initiatives either mature or quietly die. Not because the models fail, but because the operational environment refuses to tolerate their needs. The successful banks are not the ones with the flashiest algorithms. They are the ones that respect the core system's role as the keeper of truth, build modern decision layers that don't treat the mainframe as a feature database, and design for the reality that things will be late, broken, or misconfigured someday.

When the generations cooperate, something powerful happens: the bank gets both stability and adaptability. The core keeps the ledger clean. The AI layer keeps the decisions sharp. And the customer experiences a bank that feels fast without being flimsy—which, in finance, is the rarest kind of modern.

2.4 Features and Data Leakage: How Models Cheat

A banking model can be wrong in two very different ways. The first kind is the honest kind: the world is noisy, customers are complicated, recessions happen, and your model makes an imperfect guess. The second kind of wrong is more embarrassing: the model was never really guessing. It was reading tomorrow's newspaper while pretending to forecast the weather.

That second failure mode is *target leakage*. It's the single most common reason a model looks like a genius in the lab and then turns into a confused intern the moment it meets production. In plain English: the model is trained using information that would not exist at the time the bank needs to make the decision. It's like underwriting a loan using the fact that the borrower *later* defaulted. Great accuracy. Terrible time machine.

Banks are especially vulnerable because they're obsessed (correctly) with outcomes: did the loan default, did the customer churn, did the card transaction become fraud, did the collection strategy work. Those outcomes create a rich trail of timestamps, status codes, and operational breadcrumbs. Mixed carelessly into a feature set, those breadcrumbs become a secret answer key. And models, like clever students, will absolutely use it.

The ``answer key" problem, with bank-flavored examples

A feature is just an input variable: something the model is allowed to look at before it makes a prediction. The target is what you're trying to predict: default, fraud, delinquency, early payoff, complaint escalation—whatever your business cares about. Leakage happens when your features accidentally contain the target itself, or something tightly downstream of it.

That can be blatant:

- Predicting charge-off using a feature called `charged_off_flag`.
- Predicting fraud using `case_opened_indicator`.

But real leakage is usually more subtle and therefore more dangerous. It hides inside "reasonable" operational data:

- *Collections intensity as a feature.* If an account is already in trouble, the bank may start calling, sending letters, restricting credit lines, or offering hardship plans. Those actions are *responses* to emerging delinquency. If you include "number of outbound collection calls in last 7 days" to predict "probability of delinquency," you've smuggled the bank's own judgment into the model. The model will learn, essentially, "if the bank is calling a lot, it must be bad." That's not insight; it's echo.
- *"Days since last payment" computed at the wrong moment.* If your training dataset is assembled at month-end, "days since last payment" may incorporate information that wasn't available at the moment you intend to score customers (say, at application time or at statement generation time). The math is simple; the timestamp discipline is not.
- *Dispute/chargeback codes.* Fraud often results in disputes, reversals, and chargebacks. If those codes appear in the feature window that supposedly precedes the fraud decision, your model is quietly reading the post-mortem.
- *Manual review outcomes.* Many fraud systems send borderline transactions to human analysts. If the model is trained on features that include "sent to manual review" or "analyst queue ID," the model learns your process rather than the underlying risk.

The model doesn't know what is "fair" information. It only knows what reduces error. If you give it tomorrow's clue, it will take it, and it will look brilliant doing so.

What ``target leakage'' actually means in time

The easiest way to spot leakage is to stop thinking of a dataset as a rectangle (rows and columns) and start thinking of it as a timeline. Banking decisions happen at a specific moment: approve/decline, step-up authentication, raise/lower a limit, route to collections, propose a retention offer. Call that moment $T0$. At $T0$, the bank must decide using only the information available *at or before* $T0$.

Leakage is any feature that sneaks in from $T+1, T+7$, or $T+90$. It's future information. It might be a literal future timestamp ("date account closed"), or it might be a derived variable whose computation quietly uses future events ("number of late fees assessed," when late fees are assessed only after lateness occurs).

Right next to this idea, it helps to see the forbidden zone clearly:

```
# visual_id: 9z8y7x6w5v4u3t2s1r0q9p8o7n6m5l
# priority: must
# placement_hint: Next to the explanation of 'Target
# style_notes: Simple timeline with icons. Highlight
# description: A timeline chart showing a prediction

import numpy as np
import matplotlib.pyplot as plt
from matplotlib.patches import Rectangle

# Timeline coordinates
xmin, xmax = -3, 6
y = 0.0

fig, ax = plt.subplots(figsize=(7.2, 7.2))

# Base timeline
ax.plot([xmin, xmax], [y, y], color="black", linewidt

# Forbidden zone (after T0)
```

```
forbidden = Rectangle((0, -0.55), xmax - 0, 1.1, facecol
ax.add_patch(forbidden)
ax.text((0 + xmax) / 2, 0.42, "FORBIDDEN ZONE:\nFuture d
        ha="center", va="center", fontsize=11, color="#8

# Allowed zone (before T0)
allowed = Rectangle((xmin, -0.55), 0 - xmin, 1.1, faceco
ax.add_patch(allowed)
ax.text((xmin + 0) / 2, 0.42, "ALLOWED:\nOnly what exist
        ha="center", va="center", fontsize=11, color="#1

# Markers for time points
ax.scatter([0], [y], s=220, color="#1f77b4", zorder=5)
ax.text(0, -0.22, "T0\nPrediction / decision", ha="cente

ax.scatter([1], [y], s=220, color="#d62728", zorder=5)
ax.text(1, -0.22, "T+1\nLeakage event\n(outcome/process)

# Small icons (simple markers) to imply "data" and "stam
ax.scatter([-2, -1, -0.5], [0.0, 0.0, 0.0], s=[90, 90, 9
ax.text(-2, 0.18, "income\nhistory", ha="center", fontsi
ax.text(-1, 0.18, "balance\ntrend", ha="center", fontsiz
ax.text(-0.5, 0.18, "device /\nbehavior", ha="center", f

ax.scatter([2, 3, 4.5], [0.0, 0.0, 0.0], s=[90, 90, 90],
ax.text(2, 0.18, "late fee\nassessed", ha="center", font
ax.text(3, 0.18, "case\nopened", ha="center", fontsize=9
ax.text(4.5, 0.18, "chargeback /\nwrite-off", ha="center

# Styling
ax.set_xlim(xmin, xmax)
ax.set_ylim(-1.0, 1.0)
ax.set_yticks([])
ax.set_xticks([xmin, -2, -1, 0, 1, 2, 3, 4, 5, xmax])
```

```
ax.set_xticklabels(["", "T-2", "T-1", "T0", "T+1", "T+
ax.set_title("Target Leakage: Using T+1 Data to Predic

# Emphasize the boundary at T0
ax.axvline(0, color="black", linewidth=1.5, linestyle=
ax.text(0.05, 0.78, "Decision boundary", ha="left", va

# Clean frame
for spine in ["top", "right", "left"]:
    ax.spines[spine].set_visible(False)

plt.tight_layout()
```

The picture is almost annoyingly simple, which is the point. Most leakage disasters happen because the team never drew the line. People talk about "feature engineering" as if it's artisanal woodworking; in a bank, it's closer to food safety. If the ingredient didn't exist at *T*0, it's contaminated. No amount of clever modeling disinfects it.

Why leakage feels so good (and ships so badly)

Leakage has a signature smell: validation metrics that are too good to be true. AUCs that look like physics broke. Confusion matrices that read like a fairy tale. The business gets excited, budgets get approved, someone makes a deck that says "transformational uplift," and the model hits production only to perform like it's forgotten everything.

This happens because leakage doesn't just nudge performance; it inflates it. When the model sees the answer key, it doesn't need to learn real causal signals like income stability, spending volatility, merchant mix, or repayment habits. It learns shortcuts. Shortcuts are brittle.

A legitimate model is like a good underwriter: it can explain itself using stable patterns that are available at decision time. A leaky model is like a gambler with inside information: profitable until the information disappears. Production is where it disappears.

Banks experience the crash in a few familiar forms:

- *The approval model that can't approve.* In testing, it separated good and bad borrowers beautifully. In production, all the scores cluster in the middle, because the "future status code" it relied on doesn't exist yet.
- *The fraud model that misses the obvious.* In the lab, it "detected" fraud largely by learning which transactions later became chargebacks. In real time, before the chargeback exists, it loses its superpower.
- *The collections prioritization model that amplifies bias.* If the model is trained with features that include the bank's prior collection actions, it learns to replicate those actions. Historical over-collection in certain segments becomes "validated" by the model's apparent accuracy.

The most painful part is reputational. When a model collapses, the organization often concludes "AI is hype." The real lesson is narrower: *time discipline is non-negotiable.*

Leakage's sneakiest hiding place: preprocessing

Target leakage isn't always "you included a future column." Often it's procedural: the way the dataset is split, scaled, selected, or cleaned.

Here's the classic trap. A data scientist does something perfectly reasonable: fill missing values using the dataset mean, standardize variables, select the top features, maybe learn embeddings or reduce dimensionality. Then they split the dataset into training and test sets and evaluate. The model looks great.

The problem is that the preprocessing steps were *fit* on the full dataset, including the test set. That means information from the "future" (the held-out evaluation data) leaked into training. It's not temporal leakage; it's evaluation leakage. The effect is the same: your test set is no longer a clean simulation of novel data.

This is why modern ML tooling emphasizes pipelines: you split first, then fit preprocessing on training only, then apply it to test. It sounds pedantic until you've watched a bank spend six months rebuilding trust because a "simple scaling step" quietly inflated performance.

In a bank, where model risk management expects you to demonstrate performance on data not used in development, this kind of leakage isn't just a technical mistake. It can become a governance issue. Validation teams will (and should) ask: *what exactly did the model see, and when did it see it?*

The most common ``bank leakage'' patterns

Leakage is rarely labeled. It usually arrives wearing a name tag that says "helpful operational feature." A few patterns show up again and again in lending and fraud shops.

1) Outcome-adjacent timestamps

Anything that happens after the decision is suspicious, even if it feels like "just metadata." Examples:

- `loan_closed_date` used to predict default at origination.
- `first_delinquency_date` used to predict delinquency risk.
- `account_status_as_of_extract` when the extract is taken after the window you claim to be modeling.

These fields often sneak in because they're convenient keys for assembling datasets. Convenience is not innocence.

2) Post-decision fees and penalties

Late fees, overlimit fees, penalty APR triggers, and returned-payment fees are downstream signals. If the question is "will this customer go delinquent," and your features include "late fees assessed in next cycle," the model is basically grading itself after the exam.

3) Human-in-the-loop signals

Banks love "manual review" because it adds control. Models love it because it encodes a human verdict. Features like "was sent to manual review" or "case priority code" can become a proxy for the target. Worse, they create circularity: the model learns the historical workflow, then drives the future workflow, which then becomes the training data.

4) Product operations and remediation

If a customer is risky, the bank may reduce limits, freeze cards, place holds, or route the account to a special servicing unit. Those interventions can leak the bank's internal assessment into the feature set. The model becomes a mirror.

5) Aggregations computed with the wrong window

A subtle and deadly kind of cheating is using a lookback window that quietly extends beyond $T0$. For example, "average balance over the last 30 days" computed as of month-end when $T0$ is mid-month. The variable name looks fine. The computation is not.

How to stop leakage before it ships: practical disciplines

Leakage prevention is less about heroics and more about habits. The best teams treat it like operational risk: prevent, detect, and document.

Draw the scoring moment in ink

Every model should have a written definition of the scoring moment:

- What triggers scoring?
- What is the exact timestamp?
- What systems are available at that instant?
- What is the latency budget?

If this sounds like engineering instead of data science, good. Leakage often begins as a misunderstanding between "when the model is trained" and "when the model is used."

Create "as-of" feature views

A robust pattern is to build features using *as-of joins*: when pulling data from multiple tables, each join must respect the scoring timestamp. You don't join "customer profile" to "loan outcomes" in the abstract; you join the customer profile *as of* $T0$. Many banks implement this as a feature store or curated feature layer with time-aware semantics.

The practical payoff is enormous: analysts stop re-inventing brittle dataset logic, and leakage becomes harder to introduce accidentally.

Split by time, not by row

If you randomly split loan records into train/test, you can accidentally create a world where the model learns patterns from future vintages. In lending and fraud, where behavior changes with economic cycles, product changes, and policy updates, a chronological split is often the closest thing to reality: train on earlier months, test on later months.

This doesn't eliminate leakage, but it makes "future information" harder to smuggle in. It also gives you a more honest picture of degradation.

Use pipelines so preprocessing can't cheat

Every transformation that learns from data (scaling, imputation, encoding, feature selection) must be fit on training data only. The easiest way to enforce that is to make it structurally impossible to do otherwise: use a pipeline object (or the equivalent in your stack) that encapsulates fit/transform correctly.

The cultural benefit matters as much as the technical one. When teams standardize this pattern, leakage stops being a personal virtue ("I'm careful") and becomes a system property ("the workflow prevents mistakes").

Maintain a "banned features" list

A surprisingly effective governance artifact is a short, living list of feature families that require special scrutiny or are outright prohibited unless explicitly approved. In banks, the usual suspects include:

- Collections actions (calls, letters, promises-to-pay)
- Dispute/chargeback lifecycle fields
- Case management and manual review outcomes

- Status codes populated by downstream servicing
- Any fee assessed after $T0$

The list shouldn't be performative. It should be tied to code review and model validation checklists.

Make the model train like it serves

A model that trains on a rich, carefully cleaned warehouse extract but serves on a thinner, messier real-time feed is a model that will disappoint you. Some of that disappointment is "training-serving skew" (mismatched pipelines), but it often overlaps with leakage: the training data may contain fields that are populated only later, after batch jobs run, reconciliations complete, or humans intervene.

A tough but effective practice is to log the exact feature values used at serving time, then use those logged features to build future training sets. It's hard to cheat when your training data is literally what production had in hand at $T0$.

Leakage detection as an audit, not a vibe

Most leakage reviews are informal: someone scrolls through column names and shrugs. That approach fails because leakage is rarely named honestly. The best reviews feel more like an audit: systematic, skeptical, and timestamp-obsessed.

A practical leakage audit for a bank model asks questions like:

- For each feature, what system produces it, and on what schedule?
- What is the earliest time the value becomes stable?
- Does the feature rely on a batch job that completes after the scoring moment?

- If the model were run in real time, would this feature be null, stale, or different?

A nice trick: deliberately simulate production constraints. If a feature arrives via an overnight batch, set it to null for same-day scoring in a sandbox and see how performance changes. Models built on leakage often collapse immediately under this stress test. Honest models degrade more gracefully.

There's also a growing class of automated checks that scan ML code for leakage patterns—for example, spotting when a preprocessing step is fit on the full dataset rather than within a cross-validation fold. The deeper point isn't the tooling; it's the mindset shift. Leakage prevention can be treated like software quality: enforceable, testable, repeatable.

The ethics trap: leakage that launders past decisions

Some leakage isn't just about performance. It's about responsibility.

Imagine a credit risk model trained with a feature like "was previously declined by underwriting." That feature will be highly predictive of future default (declined applicants are, by definition, riskier), and the model will look strong. But what did the model learn? It learned to agree with yesterday's decisions.

If yesterday's decisions were biased, inconsistent, or overly conservative for certain groups, the model turns that history into math and calls it objectivity. Leakage can become a mechanism for institutional memory—including the parts you'd rather not preserve.

The fix is not to pretend the past didn't happen. It's to be explicit about what constitutes a legitimate predictor versus a proxy for previous judgment. Features that encode prior decisions, remediation, or operational treatment deserve special skepticism because they can create self-fulfilling loops: the bank treats customers as risky, the model learns that treatment, and the next system treats them as risky even earlier.

A bank that takes model risk seriously has to treat *process variables* as hazardous material. Sometimes they're useful (for operational forecasting, queue management, staffing). For customer-level risk predictions, they can quietly turn the model into a compliance headache.

A simple checklist that catches most leakage

Leakage hunting doesn't require mysticism. It requires asking a few stubborn questions until the story becomes consistent.

For any candidate feature:

1. *Could this value have been known at the exact decision moment?* Not "could it be computed in principle," but "did the bank actually have it then."
2. *Is the feature influenced by the target?* If the target (default, fraud, delinquency) can cause the feature to change, the arrow of causality points the wrong way.
3. *Is it a record of the bank reacting?* Collections actions, case creation, limit decreases, holds—those are the bank's moves, often based on signals you're trying to predict.
4. *Does it exist only because of an investigation or exception process?* If it appears after something went wrong, it's likely from $T + 1$.
5. *Does the feature pipeline exactly match production?* If training uses a richer dataset than serving, you may be training on fields that never exist when you need them.

Most leaky features fail within the first two questions. The last three catch the more sophisticated failures—the ones that look legitimate until you trace where they come from and when they become available.

Why the best anti-leakage tool is humility

A final uncomfortable truth: leakage is often introduced by smart people doing reasonable things under deadline pressure. The dataset is late. The business wants results. Someone adds a column "because it improves AUC." The model looks great. Everyone relaxes.

Healthy teams develop a reflexive suspicion of easy wins. If a single feature delivers a magical jump in performance, treat it like a surprise profit in a trading book: investigate before celebrating. In banks, "too good" is a risk indicator.

The reward for this paranoia is not merely a cleaner model. It's a model you can defend. When performance dips, you can diagnose it without discovering that the model's "intelligence" was an accounting artifact. When regulators or auditors ask how the model makes decisions, you can point to legitimate, time-consistent signals rather than a disguised outcome.

A model that can't cheat is forced to learn the world as it is at $T0$. That's the only world your bank gets to live in.

Chapter 3

Revolutionising Lending and Credit Risk

For decades, obtaining a loan meant fitting into a rigid box defined by a three-digit score and a manual checklist. Today, algorithms are rewriting those rules, detecting subtle patterns in data that human underwriters-and traditional scorecards-often miss. But as we hand the keys to the vault over to machines, we face a critical paradox: how do we harness the precision of AI without amplifying the biases of the past? We will unpack the mechanics of modern underwriting, the promise of invisible prime borrowers, and the regulatory tightrope of explaining exactly why the computer said 'no.'

3.1 From Traditional Scoring to ML Underwriting

A classic credit scorecard is comforting for the same reason a paper checklist is comforting: you can see the boxes. A borrower shows up as a handful of familiar facts—income, time on the job, number of late payments, credit utilization—and each fact quietly nudges the decision in one direction. More stability? Add points. More leverage? Subtract points. At the end, the ledger of points becomes a single number that feels like a verdict: approve, decline, or push to manual review.

That simplicity is not an accident; it is an operating philosophy. Traditional underwriting was built for a world where data arrived slowly, computing was expensive, and decisions needed to be explained to humans who had to live with them. A scorecard is the credit equivalent of a well-run kitchen: everything has its place, and the recipe is designed to be repeatable even when the chef changes. If a bank wants to tighten credit, it can raise a cutoff. If it wants to grow, it can lower it. The machine is predictable.

The catch is that predictability is not the same as truth. Borrowers do not behave like a tidy row in a spreadsheet, and risk does not always add up like arithmetic. Some people can carry high utilization without trouble because their income is volatile but strong. Others look pristine on paper until a single stressor arrives and the entire structure buckles. The real world is full of *interaction effects*: one factor changes the meaning of another. A long employment history can be reassuring, unless the person is in an industry being automated. A thin credit file can be benign for a high-savings borrower, but dangerous for someone living paycheck to paycheck. Linear scorecards struggle with that kind of nuance because they are designed to treat each variable like an independent knob.

Non-linear machine learning models do not treat knobs independently. They behave more like an ecosystem: the value of one signal depends on the presence of others, and the same borrower can be interpreted differently depending on context. This is where the underwriting conversation changes from *adding points* to *recognising patterns*. Instead of saying, "High debt-to-income is bad," a model might learn something closer to, "High debt-to-income is sometimes bad, but it is *especially* bad when revolving balances are rising, liquidity is falling, and the borrower's income is seasonal." That is not a single rule; it is a shape in a high-dimensional space.

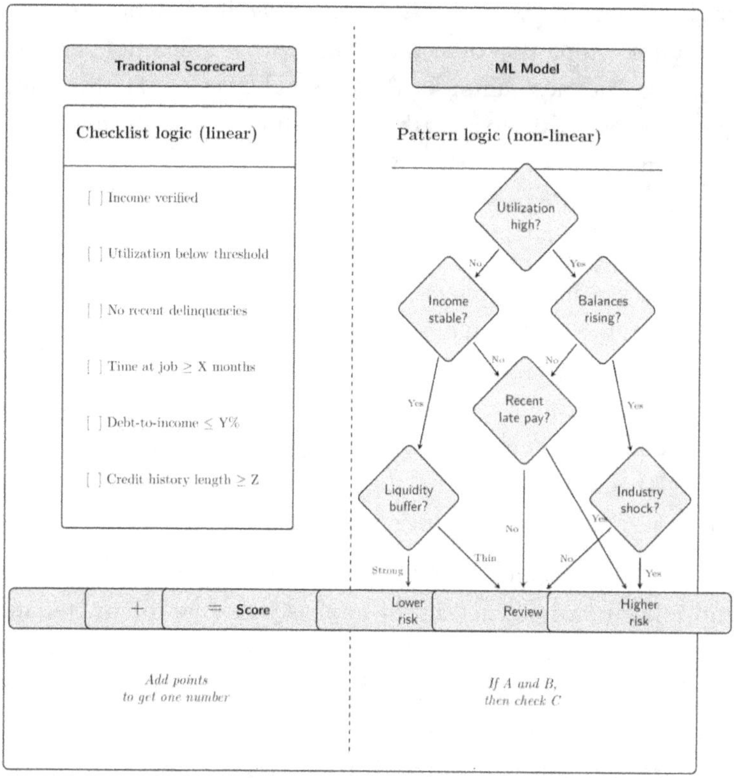

The old scorecard worldview is linear in two different ways. First, it is linear mathematically: each variable usually has a weight, and the final score is a weighted sum. Second, it is linear operationally: data comes in, a score comes out, and the score drives a decision. That pipeline is efficient, but it also encourages a subtle kind of denial: the bank acts as if its own decision boundary is a crisp line in the sand, when in reality it is more like a coastline. A coastline has bays and jagged edges; it bends around coves of safety and spikes outward into pockets of danger. Linear models draw straight borders because straight borders are easy.

To understand what machine learning changes, it helps to name the job underwriting is actually doing. A bank is not primarily trying to predict whether a borrower is a *good person*. It is trying to estimate a handful of financial outcomes, each with a different operational meaning. The most famous is *Probability of Default* (PD): the chance that a borrower fails to meet their obligations over a given time horizon. Think of PD as a dashboard gauge. If it rises, the bank should slow down: tighten credit, raise price, shorten terms, or require more verification. If it falls, the bank can responsibly take more risk.

Once you see underwriting as outcome estimation rather than moral judgment, the attraction of machine learning becomes obvious. PD is rarely driven by one variable. It is driven by combinations: income relative to obligations, liquidity relative to volatility, credit behavior relative to life events, and the difference between what a borrower *has done* historically and what they are *starting to do* now. Those combinations are where non-linear models shine.

Traditional scorecards often trace their lineage to logistic regression and related "generalized linear models." Despite the intimidating name, the philosophy is straightforward: assume the log-odds of default can be explained by adding up the contributions of each variable. If utilization rises by 10 points, increase risk by X. If time on file is short, increase risk by Y. There can be exceptions—you can add interaction terms by hand—but that is the key phrase: *by hand*. Someone has to guess which interactions matter, test them, and then decide which ones to keep. Credit teams become gardeners pruning a hedge into a shape they can maintain.

Modern ML underwriting tools like gradient-boosted decision trees flip that workflow. Instead of asking a human to pre-specify which interactions might matter, the model can discover them during training. Decision trees are built from a sequence of questions: is utilization above a threshold? is the number of recent inquiries above a threshold? is the debt-to-income ratio above a threshold? One tree is a little crude; it tends to overfit. But an ensemble of many trees—each correcting the mistakes of the last—can become remarkably sensitive to context. It is not that the model is magical; it is that it can *try* thousands of conditional rules quickly, retain the ones that improve predictive power, and discard the rest.

The difference shows up in the kind of statements each approach can express comfortably. A linear model likes sentences shaped like "more of X is bad." A tree ensemble likes sentences shaped like "more of X is bad *only when* Y and Z are true." Underwriting is full of those "only when" clauses, and they are the clauses that get lost when everything is forced into a single additive score.

Consider a borrower with high credit utilization. In a simple scorecard, that is a red flag that pulls the score down. But utilization is not just a stress indicator; it is also a behavior indicator. Some people run high utilization because they are strained. Others do it because they are gaming rewards and paying in full. A linear model can try to split the difference by giving utilization a weight that is "right on average," but "right on average" is exactly how you misprice risk at the margin—the place where most lending decisions actually live.

A non-linear model can treat utilization as a question that branches. High utilization plus high income plus a long history of paying in full can land in one branch. High utilization plus declining deposit balances plus new delinquencies lands in another. The model does not need to *understand* why those patterns matter in a human sense; it only needs to learn that they correlate with future repayment behavior. That distinction is uncomfortable, but it is also honest. Banks have always been in the business of correlation—they simply used simpler tools that made the correlations feel like rules.

Neural networks are another step along the same road. If gradient-boosted trees are a forest of conditional questions, neural nets are a mesh of learned features: they can combine inputs into intermediate representations that may not map neatly to human categories. In consumer lending, neural networks are not always the default choice (trees often win on tabular data), but they become more common when the input isn't a neat row of numbers—for example, when underwriting incorporates text from documents, image-based identity signals, or sequential transaction behavior. When the data is richer, the model's appetite for non-linear relationships grows.

The operational payoff of non-linearity is not just "higher accuracy" in the abstract. It is less bluntness. When a scorecard is blunt, it tends to compensate by using conservative cutoffs. That protects the portfolio, but it also rejects borrowers who could have paid. A more nuanced model can sometimes do something that feels like having a better set of eyes: it can say "no" to risk that was previously missed and "yes" to safety that was previously hidden. Both matter. Catching an extra slice of bad risk reduces losses; recognizing an extra slice of good risk increases growth without cheating the bank's risk appetite.

There is, however, a reason so many credit organizations stayed with linear scorecards for so long: not nostalgia, but governance. A bank is not a machine-learning lab. It is a regulated risk-taking institution whose mistakes are paid for with real losses and real reputational damage. When a model says "approve," the bank is putting capital at risk; when it says "decline," the bank is making a decision that affects a person's life. That is why model development is only half the story. The other half is the bank's ability to *control* what it built.

That control starts with something surprisingly unglamorous: data definitions. Take "income." Is it stated income, verified income, averaged income, base salary, salary plus bonus, salary minus garnishments? Two datasets can both claim to contain "income" and yet mean different things operationally. Linear scorecards often survive messy definitions because they are forgiving; they do not contort themselves to pick up tiny inconsistencies. Non-linear ML models can be less forgiving. They are excellent at extracting signal, which means they are also excellent at extracting *weirdness*. If one channel reports income differently than another, a flexible model may learn that the reporting difference itself predicts default—not because it is economically meaningful, but because it is a proxy for channel, product, or verification rigor. That is a clever shortcut and a dangerous one.

So ML underwriting tends to push banks toward a more disciplined approach to model risk management: documenting inputs, tracking lineage, testing stability, monitoring drift, and ensuring that the model's performance holds up outside the training window. In a scorecard world, back-testing might feel like a periodic audit. In an ML world, it starts to feel like maintaining a living instrument, because the world that produces credit data changes continually. Marketing changes the mix of applicants. Fraudsters adapt. Employers lay people off. Even a well-built model can become stale if the population shifts.

One of the most practical differences between traditional and ML underwriting shows up in how the bank chooses *thresholds*. A scorecard invites a single cutoff: above this score, approve. Below it, decline. Many real-world strategies add tiers and policy rules, but the mental model is still a gate.

ML models invite a different approach: treat the model output as an estimate of PD and then decide what to do with it. That may sound like a small reframing, but it changes the conversation inside the bank. Now the debate is not "What score should we require?" but "What PD can we tolerate at this price, term, and portfolio mix?" It turns a single gate into a set of dials: pricing, limits, verification intensity, and offer structure can all be tuned to the risk estimate.

A concise way to see the shift is to compare how each paradigm translates borrower attributes into action:

Question	*Traditional scorecard mindset*	*ML underwriting mindset*
What is the model output?	A score that ranks risk	An estimate (often calibrated) of an outcome like PD
How are relationships captured?	Mostly additive; interactions are hand-crafted	Interactions and non-linearities learned at scale
How does policy plug in?	Cutoffs, tiers, and rule overlays	PD-to-action mapping (price/limit/verification)
What breaks it?	Big changes in population or process	Subtle shifts, data quirks, feedback loops, drift
What does "good" look like?	Stable, interpretable, consistent	Predictive, monitored, governable, explainable enough

That last phrase—*explainable enough*—is where the romance of ML meets the reality of banking. Credit decisioning is not a private recommendation engine; it is a system with obligations. The bank needs to be able to defend the model internally, validate it independently, and communicate its decisions externally when required. Even when the law does not demand an explanation for every detail, the institution still needs a coherent narrative for why the model is safe to use.

This is why ML underwriting in banks so often converges on methods that sit in a sweet spot: flexible enough to capture interactions, structured enough to be interrogated. Gradient-boosted trees are popular partly because they can be probed. You can measure feature importance, test monotonic constraints (for example, requiring that higher delinquencies never reduce predicted risk), and apply post-hoc explanation tools that estimate how much each input contributed to a given prediction. None of that is perfect, but it gives the bank handles to grab.

There is also a more subtle reason governance matters: underwriting models are *self-fulfilling*. When you approve someone, you create future data about repayment. When you decline them, you create silence. Over time, the model learns mostly from the people it was willing to lend to. A conservative model trains on conservative approvals and can become even more conservative. An aggressive model trains on aggressive approvals and can learn patterns that only exist in that riskier slice. This is not a bug; it is a structural feature of lending data. It means that ML underwriting must be designed with feedback loops in mind, and it often benefits from strategies like reject inference, careful champion/challenger testing, and deliberate exploration (within risk appetite) to avoid becoming trapped in its own decisions.

Another way ML changes underwriting is by making it easier to combine *static* and *dynamic* signals. Scorecards tend to love static facts: age of oldest trade, number of accounts, presence of a mortgage, and so on. ML models can incorporate trends: utilization rising over three months, payment amounts shrinking, deposit balances becoming more volatile, or the timing of expenses shifting. A borrower's risk is often less about where they are and more about where they are heading. Traditional tools can incorporate trend variables, but

ML tends to incorporate them more naturally because it can learn how trends interact with baseline levels.

That said, ML does not repeal the fundamentals of credit. A model can be breathtakingly complex and still be anchored to the same economic truths: ability to repay, willingness to repay, and resilience under stress. In practice, the best ML underwriting programs look surprisingly conservative in what they *choose* to feed the model. They may use sophisticated techniques, but they privilege variables with an intuitive link to credit risk, clean definitions, and stable measurement. The sophistication goes into how those variables are combined, monitored, and governed, not into throwing every available data point into the blender and hoping the loss function sorts it out.

When that discipline is present, ML underwriting starts to feel less like a departure from traditional banking and more like an upgrade to the bank's eyesight. A scorecard is a good pair of reading glasses: simple, reliable, and effective in the conditions it was designed for. Machine learning is closer to progressive lenses: sharper across more situations, but only if the fit is right, the measurements are accurate, and someone is checking that the prescription still matches the world outside.

The final irony is that ML underwriting can make credit decisions feel *more* human, not less. Not because the model has empathy, but because it can recognize that a borrower is not a single number. Two applicants can share the same bureau score and yet be entirely different risk stories. Linear tools often compress them into the same bucket. Non-linear tools can separate them, sometimes in ways that align better with common sense: a stable household with predictable cash flow looks different from a fragile one with the same headline score. Done well, the bank moves from judging people by a blunt proxy to assessing them by a richer portrait—and that is a rare case where better mathematics can also produce a decision that feels, to the borrower, less like a stereotype.

3.2 Expanding Access with Alternative Data

A credit bureau file is supposed to be a financial biography. For a surprising number of people, it reads more like a haiku: short, incomplete, and easy to misinterpret. They pay the rent on time, keep the lights on, show up for work, and still look like a question mark to a lender because they have not used the particular instruments that bureaus are designed to record. Credit, in the modern system, is often granted to people who already have credit. Everyone else is asked to prove they exist.

Banks have a name for this missingness, and it is more revealing than it sounds: *credit invisible*. It does not mean the person is hiding. It means the data pipeline does not look in the right places. If your financial life runs through cash, debit, prepaid cards, money orders, informal lending circles, or a thin set of accounts you rarely touch, the traditional picture of you is not *wrong* so much as *underexposed*. Underwriting teams are left squinting at a dim photo and calling the blur "risk."

There is a second, more unsettling category: people who are visible but miscast. Their bureau file exists, but it tells the wrong story—or at least an incomplete one. They might have a low score because they used credit briefly, made a few mistakes years ago, then went quiet. Or they might be new to the country, new to formal banking, or recently divorced, with a file that looks chaotic while their ability to repay is steady. These borrowers are sometimes called *invisible primes*: not invisible to the system, but invisible to the system's ability to recognize quality.

Alternative data enters like a flashlight. The phrase sounds like a marketing euphemism, but the concept is simple: information that is not typically found in nationwide credit bureau files or not customarily collected on a traditional application. In other words, signals that sit outside the familiar rails of revolving balances, installment loans, and reported delinquencies. When lenders talk about "scoring the unscoreable," they are really talking about using different evidence to answer the same old question: will the borrower pay?

The most persuasive alternative data tends to be boring in the best way. Utility payments, rent histories (when reliably captured), insurance payments, and bank-account cash-flow patterns are not flashy, but they map to the muscle memory of financial responsibility. A rent payment is a monthly stress test: can you prioritize a large, non-negotiable bill and execute it reliably? Utility bills reveal rhythm and continuity: a stable household tends to have a stable pattern. Cash-flow data reveals something bureau data often cannot: how money *moves* through a person's life, not just how debt sits on their shoulders.

This is an important distinction because many arguments about alternative data get stuck on a false binary: either a lender is using "financial" information or spying on someone's digital life. In reality, alternative data spans a spectrum. On one end are transaction-derived metrics from a bank account—income regularity, expense volatility, average end-of-month balance, frequency of overdrafts, the pattern of bill payments. On the other end are non-financial digital footprints: device metadata, browsing behavior, social signals, app usage, geolocation traces. Both are technically "alternative." They do not belong in the same ethical bucket.

The line that matters is not *new versus old*. It is *intuitive versus non-intuitive*. If a lender tells a borrower, "We looked at your rent payments," most people can connect the dots. If a lender says, "We looked at the way you filled out a form and how quickly you scrolled," the dots start to feel like a constellation drawn after the fact. That intuition gap is not just a public-relations issue; it is a risk issue. The harder it is to explain why a signal relates to repayment, the harder it is to govern the model, defend it in compliance reviews, and convince yourself you are not accidentally building a proxy for something you never meant to consider.

The inclusion promise of alternative data is easiest to see through a simple thought experiment. Imagine two applicants who both have thin bureau files. Applicant A has a steady paycheck and pays rent and utilities on time, with a bank account that rarely dips below a buffer. Applicant B has irregular deposits, frequent overdrafts, and a pattern of late bill payments, but looks identical on bureau data because neither has used much credit. A traditional model sees two unknowns. A cash-flow model sees two different financial lives.

That difference matters because the cost of being misread is not evenly distributed. If you are affluent, being denied a credit card is an inconvenience. If you are building a life, it is a tax. Credit opens doors that are not even labelled "credit": renting an apartment, paying a deposit for utilities, smoothing expenses between paychecks, buying a car to get to work. A system that defaults to "no" when it cannot see you is a system that quietly charges people for being poor in the wrong format.

Banks, for their part, are not charitable organizations. They are responding to an economic puzzle: the traditional bureau file is an excellent predictor for people who have lived inside the credit system for years, but it can be a weak lens for newcomers and cash-based households. Alternative data is attractive because it promises two outcomes at once: expand the addressable market and keep losses controlled. The phrase "expand access" sounds moral; the spreadsheet translates it into something more concrete: find borrowers who will perform but are currently priced like they will not.

The technical mechanism is straightforward. Alternative data adds features. Features create separation. Separation allows a model to split a population into groups whose future behavior differs meaningfully. When a lender talks about finding invisible primes, they are talking about the model's ability to carve out a group that looks risky or unknowable by traditional measures but behaves like a prime population once you observe the right signals.

That carving, however, comes with sharp edges. The first is consent. Many forms of alternative data are not sitting passively in a bureau file; they live behind permissions. Cash-flow underwriting often relies on access to bank account transaction data, which may be provided via data aggregators and consumer authorizations. A borrower can say yes, but the experience matters. If consent is bundled into a confusing clickstream, it is technically legal and practically dubious. If consent is clear and revocable, the lender has a better foundation—and a better relationship with the borrower.

The second edge is data quality. Bureau data is not perfect, but it is standardized. Alternative data is often messy, inconsistent across providers, and sensitive to operational quirks. Rent payments might be recorded through one property manager and invisible through another. A utility payment might be bundled with roommates. Transaction data can be miscategorized: a payroll deposit mislabeled as "transfer" can change the model's view of income stability. Even small inconsistencies can matter because ML models are attentive to patterns and can learn the shape of a dataset's errors.

The third edge is durability across economic cycles. Many alternative-data models are trained during relatively benign periods or during specific market regimes. A pattern that predicts well in a stable economy may behave differently in a downturn. Cash-flow signals, for example, can change rapidly when layoffs rise or inflation changes spending behavior. A model that has never seen a full business cycle can be like a boat that performs beautifully on a calm lake and poorly in a storm. This is not a reason to avoid alternative data; it is a reason to treat it as a risk instrument that needs stress testing, monitoring, and humility.

A less discussed challenge is that alternative data can change borrower behavior—and not always for the better. Once people learn what "counts," they adapt. If rent reporting becomes a gateway to credit, landlords and tenants may both adjust incentives, sometimes in ways that produce cleaner-looking data without cleaner financial health. If cash-flow models reward a high end-of-month balance, people may time transfers to create a cosmetic buffer. This is not fundamentally different from borrowers managing utilization before applying for a mortgage; it is just happening in a new domain. Models invite gaming. Good underwriting assumes it will happen.

There is also a cultural risk inside the bank: data abundance can create overconfidence. Alternative data feels like progress because it is new, but novelty is not a credit strategy. The lender still needs to answer unglamorous questions: Is this data stable? Is it available at scale? Does it bias against certain groups? Can we explain it to a customer without sounding like we are reading tea leaves? Can we maintain it when a vendor changes an API, a categorization scheme, or a product offering? Operational fragility is a kind of risk that does not show up in AUC metrics but shows up painfully in outages and compliance findings.

The ethical boundaries are clearest when framed as a question of *relevance*. Credit underwriting is not a general-purpose evaluation of a person's worthiness; it is supposed to be about creditworthiness. Alternative data becomes problematic when it drifts from measuring ability and willingness to repay into measuring lifestyle, identity, or vulnerability. Some digital footprints can be predictive precisely because they correlate with socioeconomic status, education, neighborhood, or access to technology. A model can be "accurate" while still being unfair, because accuracy is indifferent to the moral source of the signal.

That creates a paradox. The borrowers who could benefit most from alternative data—those outside traditional credit—are often the same borrowers who have the least power to contest or even understand how they are being evaluated. If the model is built on signals that feel alien, a denial feels arbitrary. And when decisions feel arbitrary, trust evaporates. A lender can win on predictive metrics and lose on legitimacy.

The path through that paradox tends to involve choosing alternative data with a clean narrative. Cash-flow underwriting is a good example because it maps naturally to repayment mechanics. If income arrives regularly and expenses are manageable, repayment is feasible. If an account shows chronic shortfalls, repayment will be stressed. This does not eliminate fairness concerns, but it grounds the model in something both parties can discuss without needing to pretend that phone-battery level is a character trait.

Even within cash-flow data, the best signals often have a common-sense feel. Income volatility is a real stressor. A low liquidity buffer increases the chance that a minor shock becomes a delinquency. Frequent overdrafts can indicate thin margins. A sudden increase in recurring obligations can squeeze capacity. These are not perfect proxies for repayment, but they are at least in the same universe as repayment.

A practical way to think about alternative data is to ask what it adds that bureau data struggles to capture:

What the lender wants to know	*Traditional bureau data often shows*	*Alternative data can reveal*
Ability to repay next month	Past credit usage and repayment on reported tradelines	Cash-flow sufficiency, expense volatility, liquidity buffer
Willingness to repay	History of delinquencies and collections	Bill-paying cadence outside credit (rent, utilities) when reliably captured
Resilience to shocks	Indirectly, via long histories and diversified credit	How quickly cash balances recover after dips; stability of deposits
Credit experience	Depth and age of tradelines	Responsibility without formal credit (where measurable)

The table makes the promise look clean. The real world makes it complicated, because alternative data is not a single dataset; it is a supply chain. Someone collects it, cleans it, categorizes it, and delivers it. Each step is an opportunity for error and bias. If a lender relies on a third-party dataset, it inherits third-party risk. If the dataset changes, the model changes even if the code stays the same. A bank that takes alternative data seriously treats it like any critical input: it demands lineage, documentation, and the ability to audit.

There is also a strategic question hidden inside the inclusion narrative: *who captures the value?* If better models can identify invisible primes, lenders can compete aggressively for them. That competition can be healthy—more access, better pricing—but it can also shift surplus away from borrowers if the market structure allows it. A model that identifies a borrower as safer than their bureau score suggests could justify a lower interest rate, or it could justify approving them at the same high rate the market was already charging. Technology does not choose generosity. Business strategy does.

Some of the most interesting uses of alternative data are not about approval at all, but about *how* to approve. A lender might approve a thin-file borrower but start them with a smaller limit, shorter term, or more frequent payment schedule. Or it might approve with a requirement for income verification that is triggered only when cash-flow uncertainty is high. Alternative data can help tailor the shape of credit so that the borrower is not forced into a one-size-fits-all product that is either too risky for the bank or too expensive for the customer.

This is where inclusion stops being a slogan and becomes engineering. If the bank can detect that someone's income is biweekly and expenses cluster at month-end, it can offer payment timing that reduces accidental shortfalls. If it can see that a borrower's buffer is thin but stable, it can set limits that allow useful spending without inviting a spiral. The most humane lending products often look like small operational choices. ML makes those choices scalable.

The danger is that scaling can also scale harm. Alternative data can widen the decision surface, giving the model more ways to say no. A borrower might be declined not because of a late payment on a loan, but because their cash-flow pattern looks "spiky," their account has too many transfers, or their expenses are categorized in a way the model has learned to dislike. Without careful design, the bank can create a system where the reasons for denial become less legible and therefore harder to contest. Inclusion is not just about approvals; it is also about keeping the system understandable enough that people can navigate it.

One of the smartest disciplines a lender can adopt is to separate alternative data into two buckets: *core decisioning* and *supporting evidence.*

Core decisioning data should have high quality, high stability, and a clear relationship to repayment. Supporting evidence can be used to improve verification, reduce fraud, or route applications into the right workflow, but should not silently become the reason someone is denied credit unless it meets the same standards as core data. This is not a technical rule; it is a governance rule, and it is how banks avoid letting "helpful" signals become invisible gatekeepers.

For borrowers, the hopeful version of alternative data is simple: the system learns to recognize responsibility even when it is expressed outside the credit-card ecosystem. Pay your rent. Keep your account healthy. Show steady income. Get treated accordingly. The cynical version is also simple: the system finds new ways to judge you that you cannot see and cannot influence. The difference between those futures is not the algorithm; it is the choices the bank makes about what it measures, how it obtains consent, how it tests for fairness, and how it explains outcomes.

Alternative data, at its best, is not a surveillance strategy. It is a translation strategy. It takes the evidence of a functional financial life and translates it into a language underwriting systems can understand. Done carefully, it makes the credit system less circular: you do not need a long history of credit to be recognized as creditworthy. Done carelessly, it makes the system more opaque: you become a bundle of proxies, and the proxies become your fate.

Banks that want the upside without the backlash tend to follow a simple instinct: keep the model close to the economics of repayment and keep the borrower close to the truth of what is being used. When alternative data is grounded in cash flows, bill payments, and stability—and when it is collected with clear permission and governed like a critical risk input—it can turn "invisible" into "understood." And in lending, being understood is often the first step toward being included.

3.3 Fair Lending and Adverse Action

The fastest way to lose the public's trust in AI lending is also the laziest explanation a lender can give: *because the model said so*. It sounds modern, almost scientific, and it is useless in every way that matters. A credit decision is not a movie recommendation. It changes a person's cost of money, access to housing, and ability to absorb a bad week. In lending, opacity is not a quirky product choice; it is a compliance, reputational, and ethical liability.

Banks have always used models. The difference now is that models have learned to speak in sentences humans do not naturally use. Instead of "high utilization and a recent delinquency," you get a probability. Instead of "income is too low for this payment," you get a score produced by hundreds of interacting signals. The model might be right more often, yet still fail the simplest customer question: *Why?* That question is not sentimental. In many jurisdictions, it is a legal demand wearing a polite face.

Fair lending starts with a plain idea: lenders should not treat people differently for forbidden reasons. In the United States, that principle shows up in the Equal Credit Opportunity Act (ECOA) and its implementing regulation, Regulation B (Reg B). The prohibited categories are well-known in compliance departments, but the operational challenge is always the same: it is easy to avoid explicitly using protected characteristics, and much harder to avoid using *proxies*. A model can be blind to race and still learn patterns that correlate with race through geography, employment history, housing status, language preference, or even the way data is measured.

This is why the "black box" problem is not just about interpretability aesthetics. It is about control. If a bank cannot see what is driving decisions, it cannot confidently prevent discriminatory outcomes. And if it cannot prevent discriminatory outcomes, it is not just risking regulatory action; it is risking the bank's license to be taken seriously.

The phrase *disparate impact* captures a tension that makes executives uncomfortable for understandable reasons. Disparate impact does not require a smoking-gun email or an underwriter with bad intent. It is about outcomes: a neutral policy or model can disproportionately harm a protected group. ML models are particularly good at producing disparate impact accidentally, because they are particularly good at finding predictive patterns—and the world is full of historical inequities that are predictive. A model trained on yesterday's data can become an efficient machine for reproducing yesterday's unfairness.

Nowhere does this become more concrete than in the adverse action notice. An adverse action is a negative credit decision: a denial, a counteroffer that the applicant rejects, a reduction of an existing credit line, or other actions defined by regulation. An adverse action notice is the lender's obligation to tell the applicant what happened and, crucially, *the principal reasons* for it. The point is not to write a novel. The point is to give a real person a real handle: something they can fix, contest, or at least understand.

Here is the trap: modern underwriting models can be *too smart* for the old habit of generic reason codes. If a lender sends "failed to meet internal standards" or "insufficient score" as a reason, it may feel like a tidy umbrella that covers the bank's complexity. Regulators do not love umbrellas. They want specificity. The lender is not required to explain the entire model, but it is required to identify the key factors that drove the decision. That requirement does not vanish when the model is complex; if anything, complexity raises the stakes, because it becomes easier for the notice to drift away from the truth.

The difference between "truthful" and "plausible" matters. A plausible reason is the kind of thing that sounds like it *could* have been the cause. A truthful reason is the factor that actually pushed the model over the line for this applicant, in the model the bank used, at the time the decision was made, with the data the bank actually had. That is not an English problem. It is a systems problem.

The most common misunderstanding about adverse action notices is that they require a causal explanation. They do not demand that the bank prove metaphysical causality. They require the lender to disclose the principal reason(s)—the factors that were important in the decision. That may sound like a small distinction, but it is what makes the requirement feasible. A bank does not have to explain *how* a consumer's utilization mathematically affected a score. It has to say, plainly, that utilization was one of the key factors.

The AI twist is that models often do not naturally produce the kind of reasons that fit into a human-friendly notice. A gradient-boosted tree can output a PD. A neural network can output a score. Neither outputs, by default, the sentence: "Your revolving balances are too high relative to your credit limits." So the bank builds a translation layer: a reason-code engine that converts model behavior into a small set of understandable factors. This engine is not a nice-to-have. It is the difference between a scalable ML program and an expensive science project that can't be used in production.

The translation layer has to be faithful, not performative. If the model used 300 features and the notice lists three canned reasons that have only a loose relationship to the true drivers, the bank is taking a risk that is both regulatory and operational. The model might be excellent, but the bank is misrepresenting it. That misrepresentation can also hide bias. If the real driver of denials is a subtle proxy (say, a pattern tied to neighborhood effects), a generic reason code can act like a fog machine: it makes the system look less controversial while making it harder to audit.

This is where explainability tools enter the story with a mix of promise and controversy. Methods like SHAP values (Shapley Additive Explanations) estimate, for a particular decision, how much each input feature contributed to pushing the prediction up or down relative to a baseline. The key word is *estimate*. SHAP does not magically open a model and show its soul; it provides a principled attribution under certain assumptions. That can be good enough to generate reason codes, prioritize investigations, and support governance. It can also be misused as a decorative appendix that makes a model look explainable without truly making it governable.

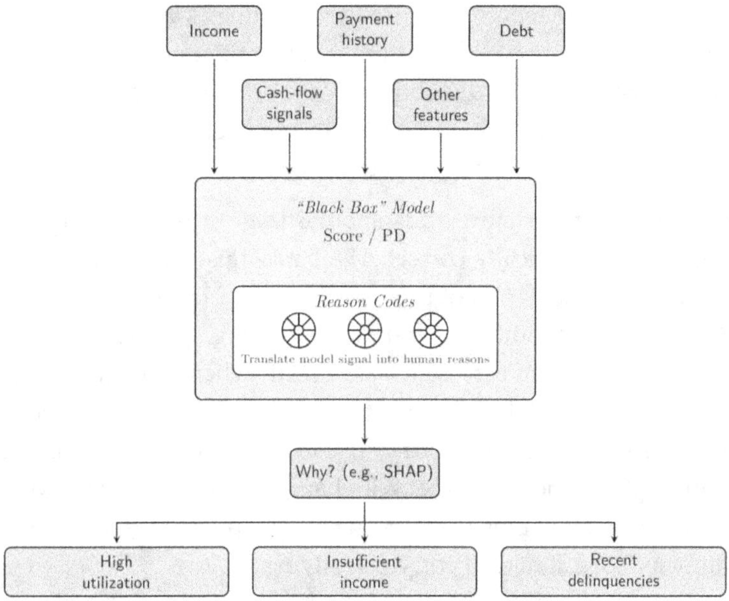

Complex models can be used in lending only if they can produce specific, defensible reasons for decisions.

There is an art to producing reasons that are both accurate and understandable. The model may rely on a highly engineered feature like "trend in end-of-month balance over 90 days." That might be a legitimate risk signal, but it is not a great reason code to send to a consumer. So the bank maps it to a more intuitive concept such as "insufficient cash reserves" or "insufficient income relative to obligations"—but only if that mapping is consistent with what the model is actually using. Otherwise the notice becomes a translation of convenience rather than a translation of truth.

This is why the most mature lenders treat reason codes as a first-class product. They version them. They test them. They validate that the top reasons coming out of the explanation layer correspond to the drivers inside the model. They monitor distributions: if the model is suddenly generating a new dominant reason, something has changed. Maybe the portfolio is shifting. Maybe a data feed broke. Maybe a new fraud pattern is contaminating the training set. Adverse action reason distributions can become a surprisingly effective monitoring dashboard because they reflect the model's *decision logic* rather than just its aggregate performance.

Fair lending governance adds another layer: it is not enough that reasons are technically correct. The bank must also make sure its model and policies do not produce prohibited discrimination. That often starts with outcomes testing: comparing approval rates, pricing, and other terms across groups, directly when demographic data is available and indirectly through proxy methodologies when it is not. The testing is rarely a one-and-done exercise. Models drift. Populations drift. The economic cycle changes. A bank that deploys ML underwriting and then stops testing is like a pilot who glances at the altimeter once at takeoff and calls it flying.

A tricky part is that ML can improve average predictive accuracy while making the tails more problematic. A model can be superb for most applicants but consistently harsher on a particular subgroup because a proxy variable behaves differently there. That is why fairness evaluation tends to look at slices: performance and outcomes by segment, by channel, by geography, by product, by time. Sometimes the biggest fairness problems hide in corners of the portfolio that look small on a dashboard.

Another source of trouble is feature selection by convenience. Alternative data and digital signals can be predictive, but their predictive power may come from correlations with protected classes or with structural disadvantages. Something as mundane as "time of application" or "device type" can encode differences in work schedules, access to technology, or income stability. A model can quietly learn that certain behaviors are associated with higher default, and the bank can end up punishing people for the circumstances of their lives rather than their repayment capacity. Even if such a feature is not illegal on its face, it can be a fairness landmine.

One practical mitigation is to impose constraints and guardrails during model development. If the bank knows that some variables are too close to protected status or too hard to justify, it can exclude them. It can also enforce monotonic relationships for certain features so the model cannot learn perverse effects (for example, that higher verified income increases risk due to an artifact in the data). Constraints are not a free lunch; they can reduce predictive power. But in banking, predictive power is not the only objective function. Compliance and legitimacy are part of the loss function whether the model recognizes them or not.

The second mitigation is to maintain a robust *human* process around the model. "Human in the loop" is sometimes sold as a moral salve. In practice, it is a control mechanism. Humans can review edge cases, handle exceptions, and spot patterns that indicate the model is drifting into nonsense. But the human layer has to be designed carefully. If manual review is applied only to certain neighborhoods or names or channels, the human layer can add bias back in. If the human reviewers use different standards than the model, the bank creates inconsistency and new fairness risks. The goal is not to replace a black box with a human box. The goal is to create a system where exceptions are governed and audited.

The third mitigation is documentation that is written like it will be read by someone skeptical and busy—because it will be. A model risk committee, an internal audit team, and regulators do not need poetry. They need clarity: what data was used, why it is appropriate, how it was cleaned, how the model was trained, how it was validated, what the limitations are, what monitoring exists, and how adverse action reasons are generated. The best documentation has an almost paranoid quality to it. It anticipates confusion and answers it before it becomes a finding.

There is a temptation to treat fairness as a constraint that slows innovation. That is a shallow reading of the problem. Fairness is also a diagnostic. When a model behaves unfairly, it often signals that the bank is relying on brittle patterns, contaminated data, or operational quirks. Fixing fairness issues can improve model robustness. A model that generalizes well across groups often generalizes better across time, too, because it is learning deeper relationships rather than shallow proxies.

Adverse action, in particular, forces a kind of honesty that many ML teams secretly benefit from. If you have to produce a reason a consumer can understand, you are less likely to build a model whose strongest signals are incomprehensible or indefensible. The requirement acts like a filter: it pushes the bank toward variables with a real-world story and away from variables that are predictive for reasons you would rather not explain out loud.

Still, honesty has a price. Once you tell borrowers why they were declined, you also tell them how to improve. That can reduce losses and increase inclusion, but it also makes gaming easier. A model that relies heavily on easily manipulated signals can be exploited. The bank has to strike a balance between transparency and resilience. The solution is rarely secrecy. It is usually to rely on signals that are harder to fake and more directly linked to repayment, and to monitor aggressively for manipulation.

The most underappreciated aspect of fair lending in ML underwriting is that it is not a one-time compliance project; it is a product discipline. Models will be retrained. Data vendors will change. Economic conditions will shift. The bank will launch new channels and new offers. Each change can bend outcomes. Fairness, like credit risk, is a portfolio property that must be managed continuously.

When it works, the system becomes more than a model. It becomes a set of promises that are kept in code and process. The bank promises that it will not discriminate, even accidentally. It promises that a denial will come with reasons that are specific enough to be meaningful. It promises that complexity will not become an excuse for shrugging. Those promises are not just about avoiding fines. They are about building a credit system that can use sophisticated pattern recognition without slipping into sophisticated unfairness—and about earning the right to say that a machine helped make the decision without turning the decision into a mystery.

3.4 Pricing, Collections, and Stress Testing

A loan is not a yes-or-no event. It is a relationship with a balance sheet attached. The underwriting decision is the opening line, but the plot unfolds over months and years: limits get raised or frozen, rates get adjusted, customers drift into hardship or climb into stability, and economic weather turns what looked like a safe portfolio into a fragile one. AI's biggest impact in lending often shows up after origination, where the bank's real job begins: *manage risk without strangling the customer experience.*

The simplest way to see the lifecycle is to imagine the bank as a thermostat. At origination, it sets the temperature: approve or decline, rate, term, and maybe a starting credit limit. After that, the thermostat keeps working. It senses changes—in the borrower, in the portfolio, in the economy—and it adjusts. Traditional lending often treated those adjustments as coarse policy moves: a quarterly tightening, a new cutoff, a refreshed scorecard. ML pushes the system toward continuous sensing and finer control, because it can update risk estimates more often and with more context.

That control starts with pricing, the least sentimental part of consumer finance. Pricing is where the bank turns risk into math and math into money. In a perfect world, the interest rate would reflect the borrower's expected loss and the bank's costs, plus a profit margin that feels fair. In the real world, pricing is also shaped by competition, customer acquisition costs, cross-sell economics, regulation, and the messy fact that borrowers react to price. Raise a rate and some customers walk away; others accept and then struggle. The bank is not pricing a static object. It is pricing a choice.

ML changes pricing by making the risk estimate less blocky. Instead of putting applicants into a few tiers—prime, near-prime, subprime—a model can output a more continuous view of risk, often framed as PD (Probability of Default), sometimes paired with LGD (Loss Given Default), the portion the bank expects to lose if default occurs after recoveries. Those acronyms can sound like an engineer's private language, but they are just gauges: PD is the likelihood of trouble; LGD is how expensive the trouble is likely to be. Together, they are the skeleton of expected loss.

With that skeleton, pricing can become more granular. A bank can decide that two borrowers who both squeak past the cutoff are not actually equal: one has a stable income and thick buffer, the other is one surprise expense away from delinquency. Traditional tier pricing might charge them the same rate because they fall in the same bucket. ML-based continuous pricing can separate them, potentially offering a better rate to the safer borrower and a more protective structure—shorter term, smaller limit, higher down payment requirement—to the riskier one.

There is a moral hazard hiding here. Granular pricing can be used to expand access responsibly, or it can be used to extract maximum willingness-to-pay from each borrower. The difference is not the model. It is governance and intent. A model that produces a finely sliced risk score is a sharp tool; it can carve better-fitting products, or it can carve the market into thinner and thinner segments until the customer's rate feels like a personalized penalty.

Even in banks with good intent, risk-based pricing has a strange side effect: it can make the portfolio more sensitive to model error. If you price in broad tiers, small mistakes wash out. If you price continuously, small mistakes turn into mispriced loans, and mispricing is contagious. Underprice risk and losses rise. Overprice risk and good borrowers self-select out, leaving you with a pool that is riskier than your model expects. This is one reason pricing models must be monitored not just for predictive accuracy, but for *selection effects*—the way pricing changes who accepts the offer.

That monitoring is easier to appreciate with a non-technical truth: borrowers are not randomly assigned to loans. They choose. If a bank's model is too optimistic and offers a rate that is unusually cheap for a risky segment, that segment will show up in volume. The model will look like it is growing the business. Then charge-offs arrive like a delayed invoice. Conversely, if the model is too pessimistic and prices safe borrowers out of the funnel, the portfolio can quietly deteriorate even as approval rates fall, because the remaining borrowers are not the ones the bank wanted.

AI also shows up in limit management, the under-discussed cousin of pricing. A credit limit is a promise: *we will let you borrow up to this amount whenever you want*. That promise has risk embedded in it, because utilization can spike quickly. Traditional limit strategies often rely on blunt heuristics: raise limits after a set number of on-time payments, cut limits after a delinquency, freeze lines in a recession. ML allows a more nuanced approach: detect changes in capacity and risk early and adjust limits proactively. This can protect the bank, but it can also protect the customer. A line reduction before a borrower is in trouble can prevent overextension. Done poorly, it can feel like a trapdoor opening under someone's feet.

The customer experience dimension matters because lending is emotional even when it is mathematical. A rate increase or a line decrease can feel like punishment, especially if it arrives without context. Banks that use ML across the lifecycle quickly learn that good risk management and good communication are not enemies. They are allies. A bank that blindsides customers in the name of "risk" may reduce losses in the short term and increase churn, complaints, and reputational damage in the long term.

Transparency requirements already shape parts of this lifecycle. In the United States, when a lender uses a credit score in connection with certain decisions, there are score disclosure obligations: the consumer may receive the score, the range of possible scores, key factors affecting the score, the date, and the source. Separate from that, risk-based pricing frameworks can trigger notices when a consumer receives materially worse terms than others. The point is not to force banks to expose their entire pricing engine; it is to prevent a world where price discrimination hides behind technical complexity.

These disclosure regimes are more than paperwork. They create a design constraint: if a bank builds a model-driven lifecycle, it must maintain the plumbing to explain key drivers and document the use of scores. The paperwork becomes a kind of pressure test. A bank can deploy an ML model, but it cannot pretend the model is not part of the decision.

Then there is the part of the lifecycle everyone hopes stays small: collections. Collections is where a bank discovers whether its models were merely clever or actually robust. It is also where the bank's brand is most vulnerable. A borrower who misses a payment is not necessarily irresponsible; they may be sick, laid off, divorced, or caring for a parent. The bank's response can be predatory, supportive, or simply bureaucratic. ML can amplify any of those styles. It can help target assistance to the right people, or it can help squeeze the most desperate customers with surgical efficiency.

The most valuable collections models are not the ones that guess who will never pay again. They are the ones that identify *early warning signals*—the small behavioral shifts that often precede delinquency. A borrower who has always paid in full starts paying the minimum. Autopay is turned off. Cash balances become choppy. Expenses outpace income for a few cycles. None of these guarantee trouble, but together they can indicate rising stress. Catching that early creates options: reminders, hardship programs, payment plans, temporary deferrals, or simply different messaging that reduces friction.

This is where AI can make collections feel less like punishment and more like triage. A good system recognizes that the "right" action depends on the borrower's situation and the bank's objectives. For a customer with strong intent but short-term shock, a flexible plan can maximize recovery and preserve the relationship. For a customer showing signs of strategic nonpayment, a firmer approach may be appropriate. Traditional collections often applied the same playbook broadly because segmentation was limited. ML can segment more effectively—but it must be governed, because segmentation can also become discrimination if it correlates with protected characteristics or if it leads to unfair treatment.

Collections also lives under a thick layer of consumer protection frameworks, especially when third-party debt collectors are involved. Rules about communication frequency, harassment, validation of debts, and dispute handling are not optional. When a bank outsources collections, it does not outsource accountability. ML-driven collections strategies can create compliance risk if they optimize purely for dollars recovered and treat legal constraints as inconvenient speed bumps. A compliant AI collections program treats constraints as hard boundaries and builds optimization inside them, not around them.

There is a more subtle risk: feedback loops. If a collections model pushes certain borrowers into aggressive treatment quickly, those borrowers may default at higher rates, not because they were destined to, but because the treatment worsened the outcome. The model then learns that those borrowers were "bad," reinforcing the strategy. This is a place where causal thinking matters. Predicting delinquency is not the same as predicting the effect of an intervention. Banks that confuse the two can accidentally train models that are great at forecasting harm and poor at preventing it.

The macro layer sits above all of this like gravity: stress testing. Individual loans fail for personal reasons; portfolios fail for economic reasons. When unemployment rises, when inflation squeezes budgets, when interest rates spike, when asset prices fall, default patterns change. Correlations rise. Safe segments start behaving like risky ones. A bank that only models normal times is like an architect who designs for sunny weather.

Supervisory stress testing regimes, where regulators provide hypothetical scenarios (often including a "severely adverse" scenario), force banks to ask an uncomfortable question: *What happens to losses and capital if the world turns against us?* The scenarios are not forecasts. They are pressure. The bank maps macro variables—unemployment, GDP, house prices, interest rates—to portfolio performance: delinquency rates, charge-offs, prepayments, utilization changes, and net interest income. This is not just a regulatory box. It is the portfolio's crash test.

ML enters stress testing in two ways. First, it can help build more flexible macro-to-portfolio models, capturing non-linear effects: unemployment rising from 4% to 6% may not be twice as bad as 3% to 4%. There are tipping points. Second, ML can help segment portfolios more intelligently. A stress test that treats the entire credit card book as one blob misses the reality that different borrower segments respond differently to shocks. Thin-file borrowers may behave differently than thick-file borrowers. High-utilization revolvers may be more sensitive to inflation. Borrowers with volatile income may be more sensitive to unemployment. ML can help estimate those sensitivities, provided the bank is disciplined about overfitting and governance.

The danger is that stress testing is precisely where ML can become too confident. A flexible model can fit historical relationships beautifully and then fail when the regime changes. Economic stress is, by definition, a regime change. During calm periods, consumer behavior may follow smooth patterns. In a crisis, policy interventions, forbearance programs, stimulus checks, and behavioral shifts can rewrite those patterns. The model's job is not to predict the future perfectly. It is to estimate vulnerability conservatively enough that the bank remains solvent and operational under pressure.

This connects directly to accounting, particularly expected credit loss frameworks like CECL (Current Expected Credit Loss) in the United States. CECL requires lenders to estimate lifetime expected losses for many assets, using forward-looking information. That forces banks to embed macro views into allowance calculations. In plain English: the bank must set aside reserves today for losses it expects tomorrow, based on reasonable forecasts and disciplined modeling. ML can improve parts of this process by extracting signal from large datasets and capturing complex interactions. It can also create model risk if the bank cannot explain, validate, and control it.

Allowance estimation is not a place where "the model was accurate on average" is good enough. Over-reserving ties up capital and constrains lending. Under-reserving creates a cliff when losses arrive and reserves must be increased suddenly, often at the worst possible time. The bank needs models that are not just predictive, but stable, interpretable enough for governance, and responsive to changing conditions without whipsawing the financial statements.

Across pricing, collections, and stress testing, one theme keeps resurfacing: AI is not merely a better calculator. It is a better *sensor*. It can detect shifts earlier, segment more finely, and respond more dynamically. That dynamism is power, and power needs discipline. Without it, the bank becomes a machine that optimizes itself into a corner.

A well-run AI-powered lending lifecycle has a particular texture. Pricing is connected to risk estimates but buffered by guardrails that prevent exploitation and instability. Limit management is proactive but not capricious. Collections is predictive but humane, focused on interventions that change outcomes rather than simply forecasting them. Stress testing is sophisticated but conservative, designed to illuminate fragility rather than impress with complexity. Most of all, the system is monitored: not just loss rates and approval rates, but reason codes, segment drift, vendor changes, model stability, and the gap between what the model expects and what the world delivers.

When banks get this right, the surprise is not that AI can approve more loans. The surprise is that it can make lending *less brittle*. It can help banks take risk with more precision and more restraint at the same time. The goal is not to eliminate losses—that would mean eliminating lending. The goal is to keep the bank from learning about risk only after it becomes a delinquency report, and to manage the life of a loan with the same intelligence that was used to approve it in the first place.

Chapter 4

The Front Lines: Fraud, AML, and Financial Crime

Imagine a bank vault that opens millions of times a second, accessible from anywhere in the world. That is the reality of modern digital banking. In this chapter, we step onto the invisible battlefield where algorithms fight to distinguish a legitimate coffee purchase from an international money laundering scheme-often in less time than it takes to blink. We will explore how AI serves as the digital gatekeeper, protecting customer assets while navigating the complex regulatory landscape of global finance.

4.1 The Cat and Mouse Game: Fraud vs. Money Laundering

Banks fight two very different kinds of villains. One is loud, impatient, and transactional: *fraud*. The other is quiet, methodical, and architectural: *money laundering*. Both show up as "bad transactions," both get routed to the same operations floors, and both increasingly rely on the same AI plumbing. But they behave so differently that treating them as one problem is like using the same lock to stop both a pickpocket and an arsonist.

Fraud is about *theft*. Someone is trying to take money that isn't theirs, usually from a specific victim, on a specific day, with a specific move. A stolen card number used for an online purchase. A hacked account used to send a wire. A fake check deposited on a mobile app before the bank realizes the image was altered. Fraud has a clock running:

the criminal wants the bank's "yes" before anyone notices, and the bank wants to stop the loss before it becomes unrecoverable.

Money laundering is about *disguise*. The money often starts as real cash or real funds in a real account—but it comes from something society doesn't want to fund: drug trafficking, corruption, cybercrime, scams, sanctions evasion, tax crimes, or any number of predicate offenses. Laundering is the act of making that money look like it came from somewhere ordinary. The launderer isn't trying to win a single transaction. They're trying to win a story: a believable narrative that survives months of scrutiny and, if it goes wrong, doesn't point back to the real people in charge.

The reason these two battles are easy to confuse is that they collide constantly. The proceeds of fraud become an input to laundering. A scammer drains a victim's account, then routes the money through mule accounts, crypto off-ramps, shell companies, or overseas transfers to make it harder to claw back. Meanwhile, laundering operations use fraud-like tactics to acquire the tools they need: synthetic identities to open accounts, forged documents to pass onboarding, account takeovers to exploit "trusted" profiles. In practice, the bank sees one tangled mess of suspicious behavior. Conceptually, it helps to separate the goals: *fraud* is "stop the theft," *AML* (anti-money laundering) is "stop the cleaning."

That difference in goal changes what "winning" looks like. Fraud teams often measure success in loss rate, approval rate, and customer friction. If your model catches more fraud but declines too many real customers, you haven't won—you've just moved the pain around. AML teams measure success in a more awkward currency: regulatory compliance, investigation quality, and whether suspicious activity gets identified, documented, and reported when it should. AML is less about stopping a single payment and more about recognizing patterns that indicate crime, then producing a defensible paper trail. Fraud is a sprint; AML is a marathon run under a microscope.

One reason fraud feels like a "cat and mouse" game is that the mouse can run right now. The criminal can test the system immediately: try a small purchase, see if it goes through, then go bigger. If the transaction is declined, adjust and try again. Fraud has fast feedback, which means the adversary can iterate fast. Laundering is still cat-and-mouse, but the mouse plays chess. The launderer can afford to be patient. They can keep transfers small, spread activity across accounts, vary counterparties, and mimic legitimate business cycles. They're not looking for the one gap in your wall. They're looking for a way to live inside your city without being noticed.

To make that concrete, picture a card-not-present fraudster sitting behind a laptop. They might have a spreadsheet of stolen card numbers, a proxy network to rotate IP addresses, and a few compromised consumer accounts to use as shipping addresses. They care about speed and acceptance. They want transactions that "look normal enough" for a few seconds. Now picture a laundering operator running a network of mule accounts. Their spreadsheet isn't card numbers; it's people. Who will open accounts. Who will receive transfers. Who will withdraw cash. Who will claim the business is a cleaning service, a consulting shop, an import/export firm. Laundering isn't just hiding money. It's hiding *relationships*.

Those relationships matter because financial crime is constrained by plumbing. Banks don't just move "money" as a concept. They move messages and ledger updates through rails that have rules. Card payments have authorization messages and settlement batches. Wires have formats, intermediaries, and cutoff times. ACH has files and return windows. Checks have image exchange and exception handling. Each rail creates different opportunities for fraud and different signals for AML. Criminals study those constraints the way engineers study APIs: what fields exist, what gets validated, what gets delayed, what can be reversed, what triggers human review.

Fraud often attacks the moment of decision: *should the bank approve this right now?* That's why fraud detection lives so close to the transaction switch. The bank has to decide before the money is gone or the goods are shipped. Many fraud fights are really fights about *authentication*: is this actually the customer? Was the device trusted? Is the behavior consistent? Does the merchant look risky? A fraud system is frequently trying to predict a short, sharp event: "this payment is unauthorized." It's an inference problem under time pressure, with incomplete data and a clever opponent.

AML, by contrast, attacks the longer arc: *why is the money moving like this at all?* That pulls in context fraud teams can sometimes ignore. Who is the customer? What do they do for a living? Where do they operate? What is normal volume for this business? Who are their counterparties, and what industries are they in? Are there high-risk jurisdictions involved? Are funds being rapidly moved through accounts with no obvious economic purpose? AML is less "is this you?" and more "does this make sense?" It's not a single prediction; it's a structured suspicion, built from fragments.

The classic laundering textbook describes three stages: *placement*, *layering*, and *integration*. Placement is getting dirty money into the financial system: cash deposits, cash-intensive businesses, money orders, prepaid instruments. Layering is the churn: transfers designed to create distance and confusion, often involving many accounts and rapid movement. Integration is the endgame: money reappearing as apparently legitimate funds, perhaps as business revenue, investments, or real estate. Real life is messier than the textbook, but the structure is useful because it explains why AML monitoring so often seems obsessed with "weird movement." Layering is the whole point. It's not noise; it's camouflage.

Fraud has its own stages, but they're more like a heist: reconnaissance, access, execution, and cash-out. Reconnaissance is phishing, credential stuffing, social engineering, buying leaked data, or probing a bank's controls with low-dollar attempts. Access is logging in, opening an account, or getting a payment instrument. Execution is the transaction itself. Cash-out is the exit: withdrawals, gift cards, crypto, merchant refunds, or shipping goods to reshippers. Where AML sees a long narrative, fraud often sees a sequence of tactics. And because those tactics are designed to work quickly, they generate a different kind of signal: burstiness, novelty, and sharp deviations.

This is where rules-based systems start to feel like trying to stop a shape-shifter with a measuring tape. Rules are attractive because they're simple: "If transfer amount > X, alert." "If customer sends money to country Y, alert." "If five login attempts fail, lock the account." Rules are also legible, which matters in regulated environments. You can point to a threshold and defend it. But criminals learn thresholds the way water learns the shape of a riverbed. They split transactions to stay under limits. They alternate between countries. They rotate devices. They run "low and slow" patterns designed to look like ordinary life.

The deeper problem is that rules struggle with *interaction effects*. A single transfer to a new payee might be fine. A single login from a new device might be fine. A single late-night purchase might be fine. But all three together, within an hour, right after a password reset, with a shipping address change? That's a story. Rules can capture stories, but they become brittle and numerous: you end up with a thicket of if-then logic that no one can confidently explain end-to-end, and every exception creates three more exceptions. The system starts to resemble folklore: "We added that rule after the incident in 2019, but don't touch it because it will break something."

Machine learning thrives exactly where rules start to suffocate: combining many weak signals into one strong suspicion. A good fraud model can take hundreds of small hints—device fingerprint, velocity of transactions, merchant history, time since last successful login, shipping distance from billing address, typing cadence, prior disputes—and produce a single score that behaves like intuition at scale. AML models can do something similar across a different feature universe: customer risk ratings, KYC (know-your-customer) data, business type, peer comparisons, counterparty networks, and patterns over weeks and months.

But the cat-and-mouse dynamic means models don't live in a clean laboratory. They live in an adversarial environment. A fraud model that works beautifully in backtesting can get "mapped" by criminals in production: they try many attempts, observe which ones pass, and infer what the model must be rewarding or punishing. In AML, adversarial pressure is slower but just as real. Criminal organizations evolve typologies: new ways to structure deposits, new mixes of mule accounts, new uses of fintech platforms, new layers involving trade invoices or crypto services. The bank's detection logic becomes part of the environment the criminals adapt to.

The bank, meanwhile, is constrained by something criminals rarely worry about: *customer experience.* Fraud controls can't simply be "maximum security everywhere." Decline too much legitimate activity and you train customers to distrust the bank. Ask for too many step-up authentications and you push people into abandoning purchases, calling support, or switching providers. AML controls have their own version of this constraint. Over-alerting doesn't just cost analyst time; it can lead to indiscriminate de-risking (closing accounts or refusing customers who seem "too hard"), which can exclude legitimate people and businesses from the financial system. In other words, both fraud and AML can fail in the same way: by becoming so defensive that they harm the very customers they're meant to protect.

The awkward truth is that neither fraud nor AML offers the comfort of perfect accuracy. The base rate problem is brutal: out of millions of transactions, only a tiny fraction are truly fraudulent or truly suspicious. Even a model that is "right" 99.9% of the time can produce a mountain of false positives if the crime rate is 0.01%. That's why operations teams talk about alerts the way emergency rooms talk about triage. The question isn't "is this definitely crime?" The question is "is this suspicious enough to spend scarce human time on it?"

This is also why the two domains produce different styles of "evidence." Fraud labels can be relatively crisp: a customer disputes a transaction; a card is confirmed stolen; an account takeover is verified. Not always crisp, but often crisp enough to train models quickly. AML labels are notoriously slippery. A case might be escalated and documented, but never conclusively proven. A SAR (Suspicious Activity Report) might be filed, but the bank may never learn the outcome of any law enforcement investigation. A customer might be exited for risk reasons, but that doesn't become a neat "laundering confirmed" label. AML teams are often training models in a fog, using proxy labels (alerts, investigator decisions, typology matches) and being careful not to confuse "we reported it" with "it was definitely illegal."

That fog changes the culture. Fraud teams can look like product teams: optimize, A/B test, tune thresholds, watch losses, ship improvements. AML teams look more like legal-adjacent investigative units: document, corroborate, justify, follow policy, and be ready for auditors and examiners. Fraud mistakes often show up as dollars lost and angry customers. AML mistakes can show up years later as regulatory findings, enforcement actions, and reputational damage. The time horizon shifts the psychology: fraud work feels like firefighting; AML work feels like recordkeeping with real stakes.

Even the word "victim" behaves differently. In fraud, the victim is often visible and immediate: a consumer whose account was drained, a merchant hit with chargebacks, a bank that eats the loss. In money laundering, the victim is diffuse: society, the integrity of markets, the communities harmed by the underlying crimes. That diffuse victimhood is part of what makes AML hard to keep emotionally vivid. A declined fraudulent purchase can feel like a win you can point to. A laundering network disrupted might be invisible from the bank's perspective: the best outcome is often that the money *doesn't* flow, or that the bank files reports that help someone else build a case.

This difference in visibility also shapes how criminals behave when confronted. Fraudsters often disappear when blocked. They move to the next target because time is money. Launderers can be stickier. They might cultivate a relationship with the bank, maintain accounts for long periods, and try to look boring. "Boring" is a strategy: regular deposits, consistent counterparties, plausible invoices, a customer service demeanor that doesn't raise eyebrows. A fraudster wants to be invisible for minutes. A launderer wants to be invisible for years.

The overlap between the two domains is why many institutions talk about "financial crime" as a combined discipline and build shared tooling. A modern bank often wants one identity graph, one customer risk profile, one feature store, one case management system, and one way to monitor performance. The same mule account can be part of an authorized-push-payment scam today and a laundering pipeline tomorrow. The same synthetic identity can be used to steal credit and to move proceeds. From an AI perspective, the economies of scale are irresistible: shared data engineering, shared model monitoring, shared governance, shared investigation workflows.

Still, combining them conceptually can hide a crucial operational distinction: fraud is frequently about making a *decision* (approve, decline, challenge) at speed, while AML is frequently about making a *judgment* (is there reasonable suspicion?) with documentation. A fraud model is often judged by its ability to reduce losses without blocking good customers. An AML model is often judged by whether it reasonably identifies and prioritizes suspicious activity in line with a risk-based program. One is optimization under latency and friction constraints. The other is surveillance under accountability constraints.

That accountability has an unusual consequence: in AML, the *explanation* can matter almost as much as the score. If a model flags an account, someone has to write down why the activity was suspicious. The reasoning has to survive internal quality assurance, audit, and regulatory examination. That doesn't mean every AML model must be simple, but it does mean the institution must invest in making complexity usable: surfacing the top drivers, showing the pattern over time, linking related entities, and capturing the investigator's rationale in a repeatable way. A black box that produces a thousand alerts a day without telling you "what story it sees" is less a model than a machine for generating compliance debt.

Fraud systems need explanations too, but the audience is often different. The customer wants a quick, respectful reason their transaction was declined, not a dissertation on embeddings and gradient-boosted trees. The call center wants a script and a resolution path. The dispute team wants evidence for representment or chargeback decisions. Regulators may care about fairness and discrimination in automated decisions, but the day-to-day urgency is practical: fix the customer, stop the bleeding, avoid training criminals by revealing too much.

So what does "cat and mouse" look like from the bank's side? It looks like a dashboard where everything is moving at once: criminals probing controls, customers changing behavior, merchants shifting risk, new payment rails emerging, and regulators updating expectations. It looks like teams arguing over the same dial. Turn it toward more security and you reduce losses but increase friction. Turn it toward more convenience and you improve growth but invite abuse. The dial isn't just a tuning parameter; it's a statement about the bank's risk appetite and its confidence in its own detection.

It also looks like a constant battle over data. Fraud wants data at the moment of truth: device signals, session behavior, merchant metadata, authentication outcomes. AML wants data with memory: who the customer is, how they behaved over months, what their peer group looks like, and how their network connects. Both want clean identifiers, consistent timestamps, and reliable lineage. If a model's features are built on shaky joins and half-trusted fields, the model doesn't fail gracefully; it fails by producing credible-looking nonsense. In financial crime, "nonsense" is expensive: it wastes human attention, annoys customers, and can create blind spots that criminals exploit.

The temptation is to imagine criminals as geniuses and banks as lumbering bureaucracies. The truth is less cinematic and more uncomfortable: criminals are often good at *process*. They specialize. They outsource. They buy tooling. They run experiments. They recruit. They industrialize. Meanwhile, banks have a different kind of sophistication: layered controls, defense-in-depth, governance, validation, and the ability to see patterns across an enormous customer base. The contest is not intelligence versus stupidity. It's one adaptive system versus another, each with different constraints and incentives.

If you had to summarize the difference in one line, it might be this: fraud tries to steal your money; laundering tries to *borrow your legitimacy*. Fraud wants value. Laundering wants cover. Fraud is the mugging in the alley. Laundering is the forged passport that gets someone through the airport for years. Both are crimes, both use the financial system, and both increasingly collide in the same data. But the bank's response can't be one-size-fits-all. The best defenses respect the enemy's objective: stop the fast theft, and unravel the slow disguise.

4.2 The 300-Millisecond Decision

A card swipe is a tiny act of faith. The customer believes the bank will recognize them. The merchant believes the bank will pay. The bank believes it can make a correct decision fast enough that the rest of the world doesn't notice there was ever a decision to make.

"Fast enough" is not a vibe; it's an engineering budget measured in milliseconds. If the answer comes back late, the terminal times out, the customer tries again, and the merchant starts treating your bank like a flaky internet connection. So fraud prevention in card payments lives inside a narrow window where every extra check has to justify its oxygen. The bank is running a miniature trial in real time: the prosecution (risk signals) and the defense (normal customer behavior) both get heard, but only in fragments, and the judge has to rule before the receipt printer gets bored.

Start with the cast of characters, because card payments are not just "customer to bank." When you tap a card, the merchant's terminal talks to the merchant's bank (the *acquirer*), which talks through a card network to the cardholder's bank (the *issuer*). The issuer is the one that says yes or no, and it's making that call with partial information. It knows what it knows about you, but it doesn't have the merchant's entire shopping cart, your facial expression, or the story you'd tell if you had time. It has a packet: amount, merchant identifiers, a few flags, maybe a device token, maybe a cryptogram, sometimes a shipping address for e-commerce, sometimes almost nothing beyond "$42.17 at a gas station."

That packet is the crime scene. Fraudsters love it precisely because it's constrained. A constrained message means constrained defenses, which means you can game the edges. If the authorization message doesn't carry a particular detail, the issuer can't directly use it. If the issuer can't depend on a detail being present, it can't build a control that requires it. Criminals don't need to understand all of ISO 8583 to exploit this; they just need to understand that the bank is making a decision without seeing the whole world.

The best way to appreciate the pressure is to imagine the transaction as a relay race where the baton is a single request/response, and every handoff adds delay. The bank's fraud model doesn't sit at the end of the line like an all-knowing oracle. It sits in the middle of a pipeline: data arrives, gets enriched, gets scored, gets translated into a policy decision, and then gets shipped back out. A model that is "accurate" but slow is like a smoke alarm that only goes off after the house is ash.

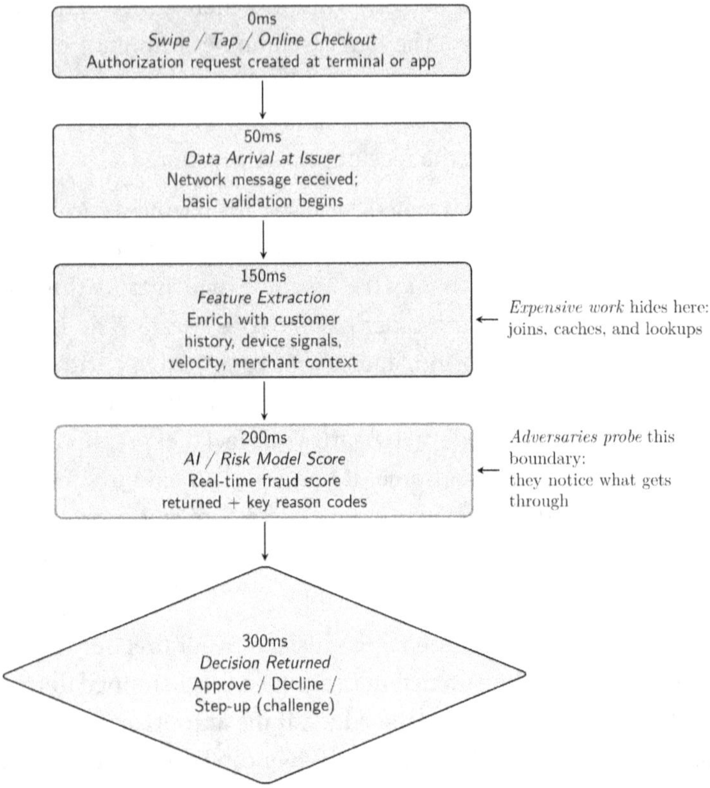

Fraud teams sometimes talk about "real time" as if it simply means "fast." In card authorization, real time means something sharper: the system must respond while the customer is still standing there, before the cashier reaches for the "try again" button, before the rideshare app cancels the trip, before the online cart evaporates. The human brain treats a few hundred milliseconds as a blink; the payments system treats it as a full season of television, packed with routing, encryption, logging, scoring, and policy.

Inside the issuer, the first layer is mundane but unforgiving: message parsing, schema validation, and routing to the right account. This is where "availability" is a fraud control all by itself. If an issuer's authorization service is down, it can't approve anything. If it falls back to a degraded mode, criminals notice. Outages and latency spikes don't just anger merchants; they create the kind of operational chaos where fraud attempts can blend into the noise.

Then comes enrichment: turning a skinny authorization request into something a model can actually reason about. That means stitching in customer history, recent transaction velocity, known devices, IP reputation, merchant behavior, geographic plausibility, authentication results (like 3-D Secure outcomes in e-commerce), and sometimes network-provided risk signals. The enrichment step is where the *AI* story becomes intensely unromantic: it's mostly data engineering under deadline. If you can't fetch the last ten minutes of activity quickly, you're asking a model to judge a movie based on a single frame.

Feature extraction is also where the bank decides what kind of intelligence it wants to be. Some features are essentially reflexes: "new merchant," "amount unusually high," "first transaction after card-not-present attempt," "many declines in a row." Others are more like memory: "typical merchant category for this customer," "usual purchase hours," "recent travel pattern." The best ones often feel almost embarrassingly simple when stated out loud. That's not a weakness; it's a sign they can be computed quickly, monitored easily, and explained without a séance.

After features, the model runs. In many production setups, the fraud score isn't a single number so much as a negotiation between systems. A model might output a probability of fraud. A policy engine might compare that probability to thresholds that depend on merchant type, amount, customer segment, or known fraud outbreaks. A second model might estimate the customer friction cost of a decline. A rules layer might veto everything if the card is reported stolen. The final decision is frequently a composite: not one brain, but an argument among specialized instincts.

That composite design is not a lack of confidence in machine learning. It's a recognition that fraud is not a purely statistical problem; it's a product, operations, and adversarial-security problem wearing a statistics costume. A model can be brilliant and still do something a bank simply can't allow, like routinely approving transactions that are likely to turn into chargebacks in a particular merchant category. Conversely, a rules layer can be "correct" in a narrow sense and still be commercially suicidal, like declining a burst of legitimate travel purchases because the customer didn't tell the bank they were flying.

If this all sounds hard, here's the part that makes it harder: the criminal is also running a pipeline. They don't see your features or your weights, but they can run experiments against your boundary. They can try a "$1" authorization at a low-risk merchant to test whether a stolen card is alive. They can vary shipping addresses, device fingerprints, and purchase categories. They can distribute attempts across botnets so velocity controls don't light up. They can run what is, in effect, a black-box audit of your model's behavior.

This is where the phrase *adversarial machine learning* stops being academic and starts being painfully literal. In the research world, "adversarial" often means tiny perturbations to images that fool a classifier. In fraud, "adversarial" means a human (or a crew) deliberately choosing inputs to get the output they want. The "perturbation" might be switching from a high-risk merchant category to a safer one, or changing the amount from "$503" to "$497" because they suspect a threshold, or spacing attempts across time zones, or forcing a mobile app flow that yields fewer device signals.

The bank's model is not only predicting fraud; it is shaping behavior. Decline patterns teach criminals which paths are blocked. Approval patterns teach them which paths are profitable. A fraud system is effectively publishing an API: send it a transaction and it returns one of three answers. That API has rate limits (velocity checks), error codes (declines), and side channels (latency, step-up prompts, soft declines). Criminals are surprisingly good at reading those side channels. If "challenged" transactions are usually still recoverable, they may avoid them. If declines correlate with certain merchant types, they'll shift. If step-up authentication is only invoked above a certain amount, they'll stay below it.

Banks try not to leak too much, but they can't avoid leaking entirely, because the system has to interact with customers and merchants. Even when the decision is opaque, the *friction* is a signal. A one-time passcode text is a signal. A "call us" message is a signal. A decline that becomes an approval on the second attempt is a signal so loud it might as well be printed on a banner.

The most underappreciated technical challenge in this millisecond world is consistency. Fraud models are trained offline on large datasets, with rich feature computation and time to spare. Production scoring is a harsher place: data arrives late, fields are missing, caches fail, upstream systems degrade, and merchants behave badly. If the model was trained on pristine features but scored on brittle ones, it will behave like a talented musician forced to play with missing strings. The output will still be music, but not the kind you want to bet customer trust on.

This is why mature fraud programs obsess over "feature availability" dashboards. Not because dashboards are fun, but because missing features are silent killers. A model can drift not only because fraud patterns change, but because the *inputs* quietly degrade. The device fingerprint vendor has an outage. A network field changes format. A new app version breaks a telemetry event. Fraud losses climb, and everyone argues about criminal innovation when the real culprit is a failed join.

Under latency constraints, engineers also make trade-offs that look strange until you live with them. You might precompute features in streaming systems so you can fetch them in microseconds during authorization. You might store "last known good" values for certain features to avoid blocking scoring when one upstream dependency hiccups. You might use simpler models in the hot path and more sophisticated models in a secondary review system, because a slightly less accurate model delivered in 20ms beats a brilliant model delivered in 400ms.

Those layered architectures are not just for speed; they're also for economics. If only a tiny fraction of transactions are suspicious, you don't want to run your most expensive analysis on everything. A cheap, fast filter can pass the vast majority of obviously normal transactions and reserve deeper scrutiny for the narrow slice that looks odd. That strategy feels almost biological: reflex first, deliberation second.

The millisecond decision also explains why fraud teams and AML teams often talk past each other even when they use similar tools. AML monitoring can afford to be slower in exchange for richer context: building cases, linking entities, tracking behavior over time, and documenting suspicion. Card authorization cannot. It must be decisive under uncertainty. The fraud model isn't trying to write a narrative; it's trying to place a bet. Approve, and you might eat a loss. Decline, and you might lose a customer. Challenge, and you might lose both if you do it clumsily.

Real-time authorization forces another uncomfortable reality: fraud prevention is not a single model, it's a *system of systems*. The model is the glamorous part, but the system includes authentication methods, customer notifications, merchant controls, dispute processes, and sometimes even education campaigns. Consider the humble push notification: "Was this you?" That's not just customer service. It's a sensor. It turns the customer into a real-time labeler, which can be more accurate than any statistical proxy. But it also introduces new attack surfaces: criminals can socially engineer victims into approving the notification, or intercept one-time codes, or convince people that the bank's warning is a scam. Every defensive move changes the game board.

The identity problem is evolving in ways that make the 300ms window feel even tighter. Traditional fraud leaned heavily on "something you have" (the card) and "something you know" (a PIN). Digital commerce blurred those lines: cards are stored in browsers, tokens sit in mobile wallets, passwords are reused, and authentication is often a prompt away. Now add synthetic identities and deepfakes to the mix. A synthetic identity is a manufactured person assembled from real and fake attributes, used to open accounts and build "legitimacy" over time. Deepfakes and voice cloning can supercharge social engineering, making it easier to trick customers, call centers, or even internal staff into taking actions that look authorized.

From the issuer's perspective, this creates a nasty paradox: the more friction you add, the more you push interactions into channels where social engineering thrives. If every unusual purchase triggers a call, criminals will start targeting the call center harder. If every login triggers a selfie, criminals will start targeting the selfie pipeline. Stronger defenses move the pressure; they don't eliminate it.

Adversarial pressure also shows up in what's sometimes called "model probing." Criminals don't need your model weights; they need to understand the boundaries. They can run many small tests to infer what you care about. Does a transaction from a new device get challenged? Does a shipping address change matter? Does 3-D Secure reduce declines? Does the bank treat a tokenized wallet payment differently from manual card entry? Over time, the criminal builds a map of the bank's reactions and chooses the path of least resistance.

Banks respond with their own counter-moves, and some of them are almost psychological. They may randomize certain controls so the boundary is harder to map. They may introduce "soft" friction that is invisible to legitimate customers but expensive for bots, like JavaScript challenges, behavioral biometrics, or device attestation. They may hold back certain decisions for post-authorization review, catching patterns that weren't visible at the individual transaction level. They may coordinate with networks and merchants to share signals that make the authorization message less skinny.

None of these moves are free. Randomization can frustrate legitimate customers who get challenged "for no reason." Behavioral signals raise privacy and transparency questions. Post-authorization review can reduce losses but might still create customer harm if the bank has to reverse transactions after the fact. Signal sharing requires governance and trust between institutions that are otherwise competitors. The 300ms window doesn't just constrain computation; it constrains ethics, customer experience, and partnership.

A final twist: criminals learn faster when you are consistent, but banks need consistency to operate safely. Operations teams need stable alert volumes. Product teams need predictable approval rates. Model risk teams need controlled changes and monitoring. Regulators expect disciplined governance. Meanwhile, fraud attacks mutate daily. The result is a delicate dance: deploy changes fast enough to keep up, but slow enough to avoid breaking production, creating bias, or accidentally declining half the cardholders in a region because a feature went missing.

This is why the best fraud teams treat their models like living things with vital signs. They watch not just loss rates, but distribution shifts in key features, latency percentiles, missingness rates, and the relationship between score bands and confirmed fraud. They compare performance across merchant categories and customer segments because criminals don't attack uniformly. They also track "shadow" outcomes: what would have happened if they had approved a declined transaction, or declined an approved one. Counterfactuals are hard, but without them you're flying by looking only at the crashes.

Underneath all the machinery, the heart of the 300-millisecond decision is a human promise: *your bank will be there when you need it, and it will not betray you when someone tries to impersonate you*. That promise has to be kept at scale, in real time, against opponents who treat your policies as puzzles and your customers as targets. The system has no luxury of certainty. It has only the luxury of speed, and even that is rationed out in milliseconds.

4.3 Uncovering Hidden Networks

A single transaction is a postcard. It tells you where it came from, where it's going, and how much it weighs. It does not tell you the relationship between the sender and receiver, who suggested the transfer, whether the sender has done this ten times today under different names, or whether the receiver is quietly acting as a drain in a larger machine. Financial crime loves that limitation. It hides in the gaps between postcards.

That's why "transaction monitoring" can become a tragic comedy when it's treated as a pile of isolated events. A rules engine sees a $9,900 cash deposit and raises an eyebrow, then sees another $9,900 deposit tomorrow and raises the same eyebrow again, as if it's encountering a new idea each time. Criminals exploit that amnesia. They don't need to outrun your fastest model; they just need to distribute their behavior across enough accounts, devices, and counterparties that any one slice looks boring.

Fraud rings and laundering networks are not just collections of bad actors. They are *coordinated systems*. Coordination is the tell. Coordination is also, conveniently, what graphs are built to represent.

A *graph* (in the data science sense) is simply nodes and edges. Nodes are entities: people, accounts, cards, merchants, devices, phone numbers, addresses, businesses, IPs. Edges are relationships: "sent money to," "logged in from," "shares device with," "received payroll from," "is director of," "delivered to." That's it. The magic comes from what happens when you stop thinking like a detective reading one witness statement and start thinking like a detective staring at a corkboard full of string.

The leap is emotional as much as technical. In the millisecond world of authorization, the bank often has to decide with partial context.

Graph thinking is an attempt to pre-build context so it's ready when the question arrives. When an authorization request hits, the bank doesn't just ask "Is this transaction odd?" It asks "Is this transaction connected to oddness?" That's a different question, and it's far closer to how criminals actually operate.

Graph analytics also changes what "suspicious" means. In classic, isolated monitoring, suspicion is often attached to the transaction: amount too large, frequency too high, geography too strange. In network-based monitoring, suspicion can attach to the *shape* of behavior. Dense clusters of accounts that transact primarily with each other. Many unrelated accounts sending funds to a single account and then going quiet. A handful of devices that seem to "birth" dozens of new customer profiles. A merchant that sits at the center of a web of chargebacks, refunds, and rapid resales. These are patterns that don't require any single transaction to scream.

Consider *smurfing*, also called *structuring*: breaking a large amount of money into smaller transactions to avoid reporting thresholds or to reduce scrutiny. In the isolated view, each deposit looks minor. In the network view, the deposits are spokes, and the target account is the hub. Even when the amounts are carefully kept below obvious trigger points, the structure betrays the intent: too many independent sources, too much synchrony, too little economic reason.

Now consider mule networks, the workhorses of modern scams. A mule is someone (witting or unwitting) whose account is used to receive and move stolen funds. Mules create distance. They also create *connective tissue* between fraud and laundering: scam proceeds become "just transfers" once they enter a mule pipeline. The bank might see a victim sending a payment that looks authorized. It might see a mule receiving funds that look like ordinary person-to-person transfers. It might see cash withdrawals that look like rent money. The criminal wants each step to look like a normal human doing normal things. Graphs are how you catch the fact that these normal-looking steps are glued together into a conveyor belt.

A helpful mental picture is the "spider in the web" structure: many peripheral accounts feed a central controller (or a small set of controllers), and money then moves outward again through different channels. The periphery exists to collect, confuse, and absorb risk. The center exists to concentrate value and make decisions. In the periphery, accounts are cheap and disposable. In the center, accounts are precious and protected.

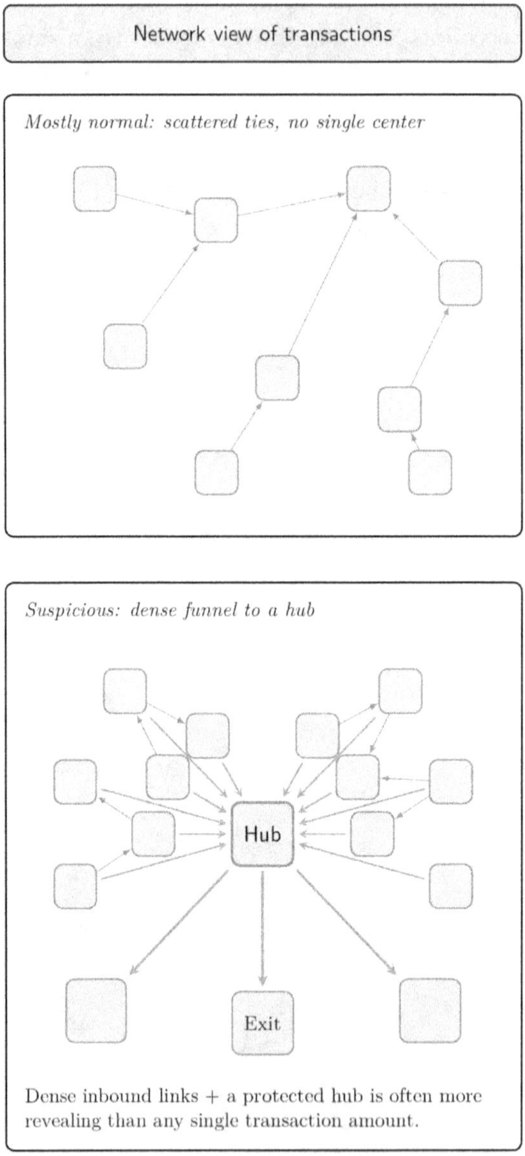

Dense inbound links + a protected hub is often more revealing than any single transaction amount.

Once you see that picture, it becomes obvious why criminals like "many small moves." In a network, many small edges can be louder than one big edge. They form a signature: a hub that shouldn't exist, or a cluster that's too tight, or a pattern of flows that looks more like plumbing than commerce.

Graph analytics also catches a subtler trick: criminals don't always funnel into a single account. Sometimes the "hub" is a *concept* spread across several accounts that act in concert. One account receives, another sends, a third withdraws, a fourth buys crypto, a fifth pays bills to make everything look domestic. The hub, in that case, is a *community*—a dense pocket of accounts with unusually strong ties. Community detection methods (ways of finding unusually connected groups) can surface this kind of "we move together" behavior even when no single account looks like the boss.

That's the real shift: moving from "Is this account risky?" to "Is this *subgraph* risky?" A subgraph is just a chunk of the network: a set of nodes and the edges connecting them. Criminal rings are subgraphs with personality. They have routines, preferred rails, common devices, shared addresses, repeated counterparties. They also have bottlenecks: places where money consolidates, where accounts get reused, where devices reappear, where a small operational team is doing a lot of work. Graphs are how you find those bottlenecks.

Even in real-time fraud, which feels like it should be too fast for network thinking, graphs sneak in through the side door. The transaction itself is an edge. The customer account is a node. The merchant is a node. The device is a node. The IP address is a node. The shipping address is a node. A "relationship-aware" fraud score is often just graph intuition disguised as features: "distance" from known fraud, "connected to" risky merchants, "shared device with" accounts that had chargebacks, "merchant connected to" many first-time buyers who immediately disputed.

This is also why criminals are obsessed with *identity fragmentation*. If they can prevent the bank from linking nodes correctly, they can prevent the bank from seeing the graph at all. A different spelling of the name. A burner phone number. A rotated device ID. A new email.

A proxy IP. A forwarding address. Each of these is a small act of sabotage against the bank's ability to build edges. In network terms, the criminal is trying to keep the graph sparse and disconnected so nothing looks like a pattern.

Banks respond with entity resolution: the unglamorous craft of deciding when two records refer to the same real-world thing. Is "Jon Smith" at this address the same as "John A. Smith" at the same address? Is this new iPhone actually the same device as last week's iPhone with a reset advertising identifier? Is this business director the same person as that personal account holder? Entity resolution can be deterministic (exact matches on strong identifiers) or probabilistic (a model that predicts "same entity" based on many weak signals). Get it right and the network becomes a clear map. Get it wrong and you either miss the ring (false negatives) or accuse the innocent by gluing unrelated people together (false positives with teeth).

A graph also forces a discipline about time. Financial crime networks are not static pictures; they are animated. Edges appear and vanish. Nodes go dormant, then revive. A mule account might be "clean" for months and then suddenly become busy for three days, then close. A scam campaign might create a burst of inbound payments to many mules, followed by rapid consolidation. That temporal texture is often the giveaway: not just *who is connected*, but *when the connections light up.*

This is one reason incremental graph methods matter in production. In the abstract, you can rebuild a graph every night. In a live financial system, "every night" can be too slow. A ring can do a lot of damage between midnight and 8 a.m. Banks increasingly want graph features that update as transactions stream in: counts of new counterparties in the last hour, number of shared devices discovered today, growth rate of a connected component (a cluster of nodes linked by edges), shortest path to known bad actors, and other "network health metrics" that change with each new edge.

The payoff is not only better detection. It's better *prioritization*. Most institutions don't suffer from a lack of suspiciousness; they suffer from an excess of it. If you generate 10,000 alerts, you haven't found 10,000 crimes. You've built a backlog. Graphs help sort the backlog by identifying the alerts that sit on top of a bigger structure. An isolated odd transaction is one thing. An odd transaction that connects two previously separate suspicious clusters is another. A single mule account is annoying. A mule account that shares a device with twelve other accounts receiving scam proceeds is a thread you can pull.

There's a human reason graphs work, too: investigators think in networks even when their tools don't. When a seasoned analyst sees suspicious activity, they immediately start asking relationship questions. Who else sends to this account? What else does this customer interact with? How did the funds move after arrival? Are there repeating counterparties? Are addresses shared? Are there round-dollar transfers that look like deliberate splitting? Graph analytics is a way of giving the investigator a head start, so their first hour isn't spent manually reconstructing what the data already knows how to connect.

That head start matters because financial crime investigations are not purely about detection. They're about explanation. A model can say "high risk," but an investigator needs to articulate a coherent pattern: the kind of pattern that can be documented, escalated, and defended. Graph visualizations and graph-derived features can turn a vague hunch into a crisp narrative: "Twenty-two unrelated accounts funneled funds into one hub within 48 hours; the hub immediately routed funds to two exits; multiple accounts share the same device and address; the flow is inconsistent with stated customer profiles." That's not just a score. That's a picture you can point to.

Graph techniques are not a silver bullet, and pretending they are is a good way to build a very expensive hallucination machine. Networks can be dense for legitimate reasons: families share addresses, roommates share IPs, small businesses share devices, gig workers have unusual payment patterns, communities send money to the same landlord, customers use the same fintech apps. A dense cluster is a clue, not a conviction. The art is in designing graph signals that capture *coordination without criminalizing community*.

That's where hybrid approaches tend to shine. Graph features feed into broader models that also consider customer context, product type, geography, and time. Some detections remain rules-based because the rule reflects a hard constraint or a clear policy (for example, certain sanctions-related restrictions). Other detections are model-based because the signal is diffuse and relational. The best systems don't worship one technique; they orchestrate them.

If fraud is often a 300-millisecond courtroom, graph analytics is the bank quietly building a case file before the trial begins. It's not about making the decision slower. It's about making the decision less blind. When criminals distribute behavior across accounts to look ordinary, the bank's counter-move is to stop staring at individual leaves and start recognizing the branch they belong to.

4.4 The Human Feedback Loop

A fraud model can spit out a score in 30 milliseconds and still be helpless, because the score isn't the decision that matters. The decision that matters is the one a human can defend at 9:12 a.m. on a Tuesday when an examiner asks, calmly and forever, "Why didn't you catch this?" The distance between those two moments is where the real work lives: queue management, evidence gathering, narrative building, escalation, documentation, and the quiet craft of turning messy reality into something a bank can sign its name to.

On paper, the handoff looks simple: model flags activity, investigator reviews, bank acts. In practice, it's a factory that can't admit it's a factory. Alerts arrive like a tide. Customers call. Merchants dispute. Internal stakeholders request "just one more rule." Meanwhile, the institution has legal obligations that don't care how understaffed the shift is. The operational side of financial crime is the place where analytics meets the human nervous system: attention, fatigue, skepticism, and the instinct to avoid making a mistake that will be replayed later.

The first surprise for outsiders is that investigators don't investigate *transactions*. They investigate *stories*. A transaction is a verb. A story has a subject, a motive, a pattern, and a reason to believe it isn't random. That's why case management tools exist at all. If the job were simply "look at a transaction and guess," you could do it in a spreadsheet. The job is "assemble enough context to justify action," which means collecting fragments: customer profile, account history, counterparties, device logs, authentication events, prior alerts, notes from the call center, even the tone of an email exchange. It's less Sherlock Holmes and more air traffic control with a legal pad.

Fraud investigations and AML investigations share a workflow shape, but they feel different under the fingers. Fraud triage is often a race against settlement, shipping, or cash-out. You freeze the card, block the transaction, call the customer, reset credentials, file a chargeback, and try to get ahead of the next attempt. The investigator's question is blunt: *Is this authorized?* AML work is slower but heavier. The question becomes: *Is there reasonable suspicion that this activity involves illicit funds or is designed to evade laws?* That question is not answered by a single yes/no label in a dataset. It's answered by assembling a pattern that can withstand review.

This is also where "false positives" stop being an abstract model metric and become a budget line. Every alert that turns out to be harmless still consumes time: minutes to open the case, minutes to gather context, minutes to document closure. Multiply that by thousands and you have a compliance organization that spends most of its life proving innocence. The irony is painful: the better you get at finding weak signals, the more weak signals you find. A naive strategy is to simply lower thresholds and celebrate higher detection. A mature strategy is to treat alert volume as a system to be engineered, not a byproduct to tolerate.

Alert volume is not just a cost problem; it's a quality problem. Overloaded analysts become pattern-blind. They skim. They learn to close quickly. They develop informal heuristics like "these merchant disputes are usually noise" or "this scenario is always junk," and those heuristics can fossilize into operational bias. Criminals don't need to defeat your model if they can hide inside the category the team is emotionally trained to ignore. In financial crime, operational fatigue is an attack surface.

That is why many institutions design detection as a layered funnel rather than a single "catch everything" net. A cheap first layer narrows the river. A more expensive layer focuses on the narrower channel. Then humans spend their time where the marginal value of human judgment is highest: ambiguous cases, novel typologies, high-risk customers, or activity that connects to other suspicious networks. The best fraud and AML organizations don't aim for "more alerts." They aim for "more *useful* alerts."

Usefulness is surprisingly concrete. A useful alert arrives with context already attached: the key contributing factors, the history slice that matters, the network connections that are unusual, and a suggested next step. A useless alert arrives naked: "Something is weird." When investigators complain about "the model," they are often complaining about *the packaging*. The model may be right. The alert may still be operationally wrong because it forces a human to do a scavenger hunt to find the why.

The term *case* is doing a lot of work here. A case is not just an alert; it is a container for evidence and decisions. A well-run case management process forces discipline: every action is recorded, every conclusion has a rationale, every escalation is traceable. That discipline matters because financial crime programs are audited, examined, and litigated in the court of hindsight. When something goes wrong, the question isn't merely "Did you have a model?" It's "Did you have a process that would have reasonably noticed and acted?" Process is the bank's memory.

At the center of that process sits a peculiar legal artifact: the *Suspicious Activity Report*, or SAR. A SAR is the bank formally telling the government, "We have reason to suspect something here." That sounds straightforward until you realize what it implies. Filing a SAR is not a prediction that a crime occurred; it is a documented assertion that the activity is suspicious enough to report. It also imposes a discipline on the institution: you can't file "vibes." You need facts, timelines, and a coherent explanation of what you saw.

SARs shape behavior even before one is filed. They influence how investigators write notes, how they think about patterns, and how cautious they are about declaring something benign. They also create a strong incentive to avoid both extremes: missing suspicious activity (the nightmare scenario) and filing low-quality reports that bury signal in noise (the slow-motion failure). The best investigators write SAR narratives the way good journalists write: clear chronology, relevant context, no melodrama, and enough specificity that someone else can act on it.

A detail that matters for AI is that investigator decisions become *labels*. Labels are the lifeblood of supervised learning: examples of "fraud" and "not fraud," "suspicious" and "cleared." In fraud, labels often arrive via disputes and chargebacks, which are noisy but plentiful. In AML, labels arrive through case outcomes and SAR decisions, which are scarcer, slower, and sometimes ambiguous. Either way, the model's future accuracy depends on the integrity of those labels.

This is where reality throws sand in the gears. An investigator's decision is not a pure reflection of truth; it is a reflection of truth filtered through policy, time pressure, training, tooling, and risk appetite. Two investigators can see the same facts and reach different outcomes. A case might be closed because the activity seems explainable, or because the queue is full, or because the analyst can't access a key system at that moment. If you treat every closure as "not suspicious" and feed it back into training without nuance, you train the model on operational constraints, not criminal reality.

Strong programs treat investigator outcomes as *structured feedback*, not a single binary switch. They capture reason codes: what made it suspicious, what evidence cleared it, what documentation was missing. They capture confidence levels. They capture "unable to conclude" as its own category. They capture the difference between "false positive" and "true positive but low risk." This extra structure costs time to collect, which is why tooling matters so much. If analysts have to type essays into free-form boxes, they will not do it consistently. If they can select meaningful, well-designed categories with minimal friction, the institution gets a dataset that can actually make models better.

Overrides are the most honest form of feedback because they are moments of disagreement. A model says "approve," an analyst says "no." Or a model says "high risk," and an analyst says "this is a payroll run; stop bothering us." Those disagreements are gold. They tell you where the model is blind, where features are missing, where policies are misaligned, or where the world has shifted. They also come with explanations, because an override must be justified to the next person who asks.

Yet overrides can also be poison if they're treated as "human knows best" in a simplistic way. Humans are inconsistent, and criminals learn human patterns too. An experienced investigator may have excellent intuition, but intuition doesn't scale, and it can be biased. If override behavior is itself biased against certain customer segments or geographies, feeding overrides back into the model can embed that bias into automation. This is one reason governance matters: not just to satisfy regulators, but to prevent the feedback loop from becoming a rumor mill that learns the wrong lessons very efficiently.

The mechanics of retraining are less glamorous than the phrase "active learning" suggests. Active learning is a family of techniques where the model identifies examples it is uncertain about (or expects to be most informative) and routes them for human review, so the resulting label improves the model faster than random sampling. In financial crime, that idea is operationally attractive because human review is expensive. If you can spend your scarce human attention on the cases that most improve the model, you get compounding returns: each investigation makes tomorrow's alerts smarter.

The catch is that "most informative for the model" is not always "most important for the bank." The model might be most uncertain about edge cases that are harmless but unusual. The bank might care more about catching a high-risk typology even if the model feels confident. So the practical version of active learning in banks is rarely pure. It is constrained by risk priorities, regulatory expectations, and operational capacity. Human feedback is steered not only by uncertainty, but by what the institution is obligated to care about.

This steering is where the term *risk-based* stops being a slogan and becomes an operating system. Banks don't have infinite investigative capacity, so they allocate effort based on customer type, product, geography, channel, and known threats. That allocation should shape both the detection logic and the learning loop. If a bank serves a large population of small businesses with cash-intensive patterns, it should not train its AML model to treat cash as inherently suspicious; it should train it to understand *which* cash patterns are inconsistent with the customer's story. If a bank offers instant person-to-person payments, it should design feedback loops that learn scam typologies quickly, because criminals love speed.

Speed is the hidden theme of the human loop, too. AML is not scored in milliseconds, but timeliness still matters. Alerts that sit untouched for weeks can become compliance failures. Fraud disputes have windows. Customer harm has momentum. The operational loop has its own latency: time from alert to review, time from review to action, time from action to label, time from label to model update, time from model update to measurable impact. A bank that retrains models quarterly while criminals iterate weekly is bringing a chess clock to a street fight.

That's why many institutions adopt "champion/challenger" setups: a champion model runs production decisions; challenger models run in parallel, scoring the same traffic but not acting on it. Challengers can be evaluated on recent data without risking customer experience. If a challenger outperforms, it can be promoted with controlled change management. This is not just technical hygiene. It's how you create learning velocity without turning customers into test subjects.

Monitoring is the other half of learning velocity. A deployed model can degrade quietly. Fraud patterns shift. Merchant behavior changes. New authentication methods alter feature distributions. A marketing campaign changes customer mix. A vendor changes how device IDs are generated. If you only look at overall fraud losses, you may notice the problem after the damage is done. Good monitoring watches early indicators: population stability (are inputs shifting?), score distribution drift (is the model suddenly more "confident"?), calibration (do predicted risks match observed outcomes?), segment performance (is it failing in a specific channel?), and operational metrics (did alert volume spike because a feature broke?).

None of this works without trust between the humans and the machines. If analysts believe the model is a random number generator, they will ignore it and build their own shadow heuristics. If modelers believe analysts are irrational obstacles, they will build systems that are technically elegant and operationally unusable. Trust is earned through small, concrete behaviors: the model surfaces the top reasons an alert fired; the alert links directly to the evidence; the investigator can give feedback without filling out a novel; the system learns from that feedback in ways analysts can observe. When people can see that their effort improves tomorrow's queue, they invest in the loop.

There is also a privacy and data-sharing tension lurking in the feedback loop. Better labels and better network context often require linking data across products, channels, and sometimes institutions. But the bank can't treat every piece of customer data as freely combinable. Responsible programs use governance: purpose limitation, access controls, audit trails, and sometimes privacy-enhancing techniques that reduce unnecessary exposure. The point isn't to hoard data; it's to use it with enough care that the institution can defend both its effectiveness and its restraint.

The human feedback loop is, ultimately, the bank admitting a humble truth: models don't "solve" financial crime. They shift the work. They triage. They surface patterns. They accelerate attention. Then humans decide what the institution will claim, what it will block, what it will report, and what it will tolerate. Every one of those human decisions becomes training data, policy, and precedent. The loop doesn't just make the AI smarter. It makes the bank *more itself*—for better or worse—because the system learns the bank's priorities and blind spots at scale.

A bank that treats investigations as a compliance afterthought gets an AI that learns paperwork. A bank that treats investigations as a craft—with tooling, training, quality control, and feedback captured in a usable form—gets an AI that learns judgment. The difference is not in the algorithm. It's in whether the organization respects the moment when a person looks at a messy situation and says, "Here's what I believe happened, and here's why I'm willing to put it in writing."

Chapter 5

Reimagining Customer Experience and Growth

Remember when banking meant walking into a branch where the manager knew your name and asked about your kids? Digital banking gave us convenience but traded away that relationship. Now, artificial intelligence offers a way to bring the 'personal' back into banking at a massive scale. In this chapter, we explore how banks are moving beyond clumsy chatbots to true digital assistants, turning distinct data points into helpful financial nudges, and listening to the 'voice of the customer' in ways human teams never could.

5.1 Conversational AI and Contact Centers

Banks did not invent the automated conversation, but they perfected the version that makes people swear at their phones. The early era of banking chatbots and IVRs (interactive voice response menus) was built like a maze: choose from a numbered list, take the wrong turn, get shoved back to the start. These systems were not *stupid* so much as literal-minded. They could do the few things they were designed for—"What is my balance?" "When did my paycheck arrive?"—and then collapsed into polite failure the moment a customer spoke like a human: *I am traveling and my card got declined twice; also I need to raise my limit for tonight.*

That gap between how customers talk and how systems were built created an awkward dynamic inside banks. Operations teams wanted containment (more issues handled without an agent). Compliance teams wanted accuracy and traceability (no improvising, no creative interpretations). Customers wanted empathy, speed, and answers that did not require memorizing a phone-tree. The compromise was the scripted bot: safe, brittle, and maddening. Scripted bots could be audited because they were essentially a set of pre-approved sentences. But the cost of that safety was the customer experience: the interaction felt less like getting help and more like trying to guess the password to a door you are already allowed to open.

Large Language Models (LLMs) changed the shape of that trade-off, not by making bots "smarter" in the way a person is smart, but by making them better at the mundane miracle of conversation: holding context. A modern model can keep track of pronouns, implied intent, and the difference between "my card *was* stolen" and "I think my card *might have been* stolen" without forcing the customer through a rigid form. It can ask clarifying questions the way a good teller would. It can paraphrase what it heard, confirm, and only then act. The interaction stops feeling like a menu and starts feeling like a dialogue.

That is the glossy part of the story. The less glossy part is that an LLM can also produce a sentence that *sounds* right and is dead wrong. In a retail bank, being plausibly wrong is worse than being obviously unhelpful. A decision-tree bot that refuses to answer is annoying; a conversational bot that casually invents a fee waiver policy is a compliance incident waiting for a case number. In other words: the evolution from scripted bots to fluid conversation does not remove risk. It *moves* risk from the customer experience layer to the governance layer. The bank gets a more human front door, and therefore needs better locks.

The contact center is where this tension becomes measurable. Every call, chat, and message is a tiny stress test: of product design, of policy clarity, of operational resilience, and of customer trust. When something breaks (an app login loop, a failed bill-pay batch, a card authorization glitch), the contact center does not just hear about it. It becomes the place where the institution either regains credibility or bleeds it. Conversational AI sits right at that pressure point, which is why the question is not "Can an AI answer questions?" The question is "Can an AI handle uncertainty without causing harm?"

A helpful way to think about the safe deployment of conversational AI in banking is *triage*. Not every interaction deserves the same treatment. Some are low-stakes and repetitive: *What is my routing number?* Some are medium-stakes but straightforward: *I forgot my password.* Some are high-stakes because they touch money movement, legal rights, fraud, or vulnerability: *I do not recognize these charges. I need to stop a payment. I am being threatened to send a wire.* The AI should not pretend those are all the same kind of conversation. A safe system behaves like a good front-desk nurse: it listens, sorts, and escalates quickly when the symptoms point to danger.

That mindset leads naturally to a human-in-the-loop design, not as an afterthought but as a core feature. The ideal hand-off is not a clunky "Please call this number" dead-end. It is a seamless baton pass: the customer does not have to repeat themselves, and the agent does not have to start from zero. The AI collects the essentials, confirms what will happen next, and then transfers context—a brief, structured summary plus key entities (dates, amounts, merchant names), plus any signals it detected (possible fraud, negative sentiment, repeated contact). Done well, it feels like the bank is one organism, not two separate species awkwardly sharing the same customer.

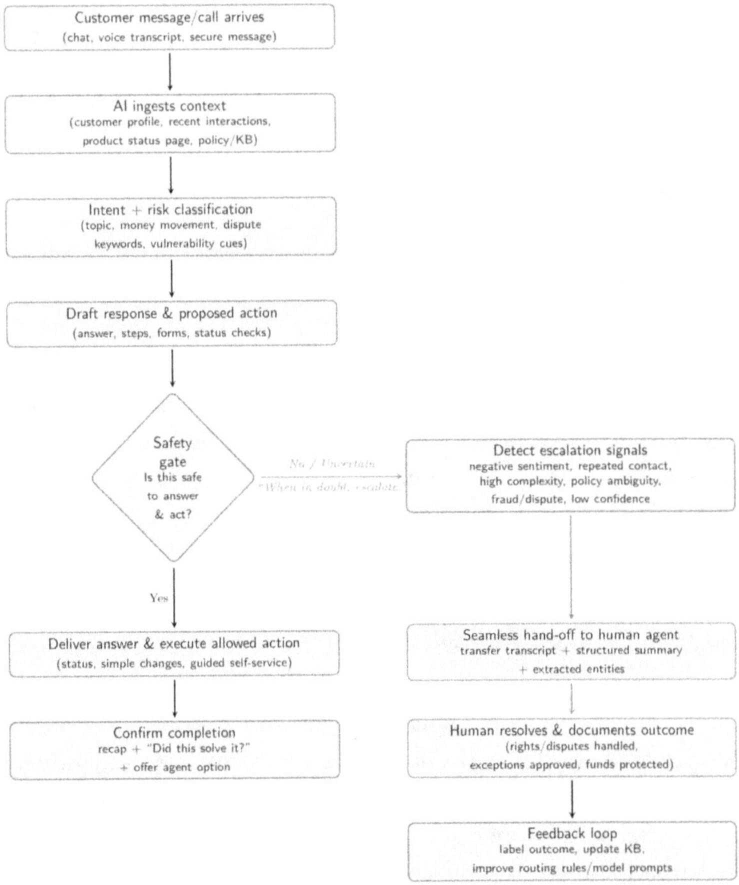

The most underappreciated ingredient in that hand-off is not the model. It is the *interface contract* between the AI and the human team. If the AI hands an agent a paragraph of mushy prose, the agent will skim, miss details, and ask the customer to repeat themselves. If it hands over a crisp bundle—"Customer reports $247.19 charge at Merchant X on 2026-02-28; card in possession; two prior declined transactions; sentiment strongly negative; likely dispute"—the agent begins the call already halfway to resolution. Good agent experiences create good customer experiences; bad agent experiences create the kind of calls that end with "I want to close my account."

That contract also determines what the AI is allowed to do. Banks often start by letting the bot *talk* but not *act*: answer FAQs, link to forms, guide navigation. Then they graduate to low-risk actions such as updating contact preferences, setting travel notices (where supported), or ordering a replacement card with clear confirmations. The seductive mistake is to jump too fast into money movement or credit decisions. The bot that can speak in full sentences makes executives feel like it can do everything in full autonomy. But operational risk does not care about eloquence. The safest AI is frequently the one that is opinionated about what it *will not* do.

Three kinds of guardrails matter most in consumer banking contact centers: factual guardrails, rights-and-policy guardrails, and privacy guardrails. Factual guardrails are about preventing the model from fabricating. That includes grounding responses in authoritative sources: product disclosures, fee schedules, rate sheets, and real-time system status. It also includes refusing to answer when the system does not have the right data. A bot that says "Your wire should arrive in 1–2 hours" because that sounds plausible is a bot that will eventually become a screenshot in a complaint.

Rights-and-policy guardrails are more subtle and more important. Customers do not call a bank only to ask for information. They call to assert rights: disputing charges, reporting errors, requesting payoff statements, asking for accommodations, escalating a complaint. A conversational system must be trained—and explicitly designed—to recognize the language of these moments. "I do not recognize this transaction" is not just "card help"; it is the start of a regulated workflow with timeframes, notices, and specific steps. "I need to stop payment" is not a suggestion engine opportunity. It is an urgent operational request with consequences. The bot should be less like a cheerful concierge and more like a careful air-traffic controller: calm, precise, and unwilling to wing it.

Privacy guardrails are the third leg. Contact centers are messy in the real world. People blurt out Social Security numbers, full card numbers, and addresses even when you tell them not to. Voice transcripts and chat logs are gold for analytics and training, which is exactly why they become risky to store and share. Privacy-safe design usually means automatic redaction of sensitive data in transcripts, strict retention policies, and tightly controlled access for anyone who reviews conversations. It also means being careful about what context the bot is allowed to retrieve. A system that can answer "What was my last transaction?" is useful; a system that can casually display every merchant and every location from the last 90 days in the middle of a shared device session is a problem.

There is a second privacy wrinkle that is less talked about: *the customer's sense of being watched.* Personalization can feel like service or surveillance depending on tone and timing. "I see you attempted a transfer that failed" can be helpful; "I noticed you bought cough medicine and searched for clinics" is creepy, even if it is technically derived from permitted data. Conversational AI heightens this sensitivity because it speaks in a human register. The bank's best defense is restraint: only mention data that is obviously relevant, expected in context, and necessary to solve the problem.

Once guardrails are in place, conversational AI becomes powerful in two distinct roles: self-service and agent-assist. Self-service is the visible chatbot: the system the customer interacts with directly. Agent-assist is the backstage co-pilot: real-time transcription, suggested knowledge articles, next-step prompts, and wrap-up summaries that cut after-call work. These two roles are often bundled under the same "AI contact center" budget line, but they behave very differently. Self-service carries brand risk because it speaks to customers. Agent-assist carries operational risk because it influences employees. Both need controls, but the failure modes differ. A self-service bot that says the wrong thing misleads a customer. An agent-assist system that suggests the wrong thing can industrialize mistakes, scaling a misunderstanding across an entire floor of agents in a single shift.

Agent-assist is also where the benefits can be surprisingly humane. Contact center work is cognitively brutal: switching systems, hunting policies, staying polite under pressure, and summarizing everything for a case record. A good co-pilot takes away the administrative drag and leaves the human with the part that humans are best at: judgment and empathy. But the bank has to engineer that outcome. If the AI floods the agent with pop-ups, low-quality suggestions, and irrelevant confidence scores, it becomes another screen to manage—a new kind of fatigue. The best implementations feel less like "AI" and more like having the right sticky note appear at the right time.

The *timing* of assistance matters as much as the content. Customers get angriest when they feel the bank is wasting their time. That anger is usually triggered not by the problem itself (systems fail; fraud happens) but by repetition: repeating identity checks, repeating the story, repeating the same troubleshooting steps. AI can cut repetition by stitching together context across channels. If a customer tried to resolve an issue in chat and then called, the system should treat it as one continuous conversation. The agent should see what the bot already collected. The bot should know that the customer already tried rebooting the app and clearing cache. This is boring plumbing work—identity linking, session handoff, state management—but it is exactly where conversational AI either becomes magical or becomes another layer of friction.

The triage approach also forces a sober conversation about *confidence*. Models can output a probability score or some internal measure of uncertainty, but a bank should not rely on the model's self-esteem as the only trigger for escalation. Some situations are simply high-stakes regardless of confidence. Others are complex even if the model is fluent. The better question is: *What is the cost of being wrong here?* For a routing number, the cost is low. For a dispute, the cost includes customer harm, regulatory exposure, and reputational damage. That is why guardrails should be policy-driven, not purely model-driven.

A practical trick many teams adopt is a small set of "red-flag" topics that always hand off to humans or to specialized workflows: fraud, disputes, debt collection-related questions, hardship and forbearance requests, legal threats, and anything involving minors or deceased account holders. These are not just complicated. They are emotionally charged and procedurally constrained. The bot's job is to recognize the red flag quickly, gather only what is necessary, and move the customer to the right place without theatrics. Customers do not need the bot to prove it can handle everything. They need the bank to prove it can handle *their* problem responsibly.

The other trigger is sentiment—not because the bank should treat angry customers differently in a moral sense, but because negative sentiment predicts escalation, repeat contacts, and churn. Sentiment detection is an imperfect instrument; sarcasm and cultural variation can fool it. Used carefully, it is less a judgment and more a smoke detector. If the model sees a conversation heating up, it can switch modes: shorter messages, clearer options, stronger confirmations, and quicker escalation. The goal is not to "calm the customer down" with synthetic empathy. The goal is to stop wasting time. Speed and clarity are the most reliable de-escalation tools a bank has.

One of the most counterintuitive realities of conversational AI is that *politeness can be dangerous*. Models are trained to be helpful, and helpfulness often expresses itself as reassurance. But banking customers do not need reassurance; they need accuracy. A bot that says "Do not worry, your payment will go through" is doing emotional labor with a promise it may not be allowed to make. A safer pattern is procedural: "I can check the status. If it is pending, here is what that means. If it failed, here are the next steps." This sounds less warm, but it is more trustworthy. In a regulated environment, trust is warmth.

That brings us to a critical design question: should the bot *sound* human? Many banks are tempted to give the bot a name, a personality, and a chipper tone. Sometimes that works for low-stakes retail interactions. But in money problems, customers tend to prefer competence over charm. A neutral, concise voice reduces the sense of being manipulated and makes it easier to insert necessary disclaimers without sounding evasive. If you want the interaction to feel human, the best path is not jokes. It is the human-in-the-loop: a fast handoff when the conversation becomes delicate, confusing, or consequential.

Under the hood, safe conversational AI depends on a few unglamorous disciplines. The first is knowledge management. If policies live in PDFs, tribal memory, and outdated intranet pages, the bot will be inconsistent because the bank is inconsistent. The second is change management. When a fee changes, a product is sunset, or a system goes down, the bot must update quickly and predictably. Otherwise, it becomes a loudspeaker for stale information. The third is monitoring: not just uptime, but conversation quality. Banks need to review transcripts, sample outcomes, measure escalations, and treat weird clusters of interactions as early warning signs. When the bot suddenly sees a surge in "I cannot deposit checks" messages, that might be an AI issue. Or it might be a mobile update that broke remote deposit capture. Either way, the bot is the sensor.

Monitoring also means watching for a unique risk: the bot that is *technically* correct but practically misleading. A disclosure can be accurate and still confuse. "Funds availability is subject to our policy" is true and useless. Conversational AI gives banks an opportunity to translate policy into plain language: "This deposit may be held for up to two business days. If it is held, you will see the availability date in your app." That translation is customer experience in its purest form. But it requires rigorous review, because plain language can accidentally add commitments the policy does not make.

The best teams treat conversational AI outputs as bank communications, not as software outputs. That is a subtle shift with big implications. Bank communications get reviewed. They have approved language for certain topics. They have escalation paths for complaints.

They are tested for readability and accessibility. When the chatbot becomes a primary channel, it becomes part of the institution's communications apparatus, with all the responsibilities that implies.

A final point that separates mature deployments from flashy demos is the willingness to constrain the problem. The dream is a single conversational agent that handles every product, every channel, every nuance. The reality is that banking complexity is not evenly distributed. A small number of intents account for a large share of volume: balance inquiries, password resets, card activation, transaction status, fee explanations. Start where the data is clean, the workflows are well-defined, and the consequences of error are low. Get the handoff experience right. Make the bot excellent at a small set of things, and unapologetically quick to escalate everything else. Customers forgive boundaries; they do not forgive confident nonsense.

Conversational AI in contact centers is not a replacement for human service so much as a redesign of the front line. It can absorb the repetitive questions that drain agents, and it can act as a high-speed receptionist that routes complex issues to the right human with the right context. But it only earns that role when the bank treats it like a regulated product: bounded authority, documented behavior, rigorous monitoring, and a safety net that works *every time*. The future of customer experience in banking will not be the most human-sounding bot. It will be the most trustworthy system—one that knows when to speak, when to ask, and when to get out of the way.

5.2 Personalisation and Recommendation Engines

Most bank marketing has the emotional intelligence of a toaster. You pay off a loan early and get congratulated with an offer for the same loan. You move apartments and receive a credit-card promo addressed to the old place. You carry a large balance for months and the bank helpfully reminds you that "saving is important." None of this is malicious; it's just what happens when messaging is built from broad segments ("young professional," "mass affluent," "small business") and blunt calendars ("send at the end of the quarter") rather than signals that a real person is doing something real.

Personalisation is the bank's attempt to stop talking like a billboard and start behaving like a helpful assistant. Not a chatty friend, not a nagging parent—more like the one colleague who quietly points out, "If you submit that expense report today, it'll clear before payroll." In banking, that idea often gets packaged as *Next Best Action* (NBA): given what the bank knows right now, what is the most useful thing to do next for this customer *and* the institution? "Action" can mean sending a message, showing an offer in the app, prompting a switch of payment method, recommending a product, or deciding to do nothing at all.

The interesting part is that the NBA frame forces banks to admit something uncomfortable: most "personalisation" efforts are just dressed-up selling. A recommendation engine can be a service layer or a pressure layer. If the model exists mainly to increase cross-sell, it will eventually learn to behave like an aggressive salesperson: interrupt, repeat, and optimize for the click rather than the relationship. If the model is designed as a service layer, it learns a different posture: *relevance* over volume, timing over frequency, clarity over persuasion. The same machine learning toolkit can build either. The difference is not the math; it's the objective function and the guardrails.

A useful mental model is to imagine the bank as a GPS for money. A GPS doesn't shout random street names at you all day. It waits until the moment a decision matters—the fork in the road—and then gives a specific instruction with enough context to trust it. "In 500 feet, keep left to stay on I-95." That's what good financial personalisation feels like: not a stream of generic financial wellness slogans, but targeted nudges at the moment a customer is already making a move.

Timing is where banks have an unfair advantage compared to most industries: money creates timestamps. A paycheck hits at a predictable cadence. Rent shows up like a metronome. A card purchase is a real-time statement of intent: groceries, travel, medical, entertainment, a last-minute hotel because your flight got canceled. Personalisation becomes plausible when the bank stops viewing transactions as an accounting ledger and starts treating them as behavioral data—carefully, ethically, and with restraint. When a customer buys an airline ticket, the bank doesn't need to scream "TRAVEL CARD!!!" It can quietly do something more useful: surface the card that has no foreign transaction fees, remind the customer to set a travel notice if needed, or offer an instant virtual card number for booking protection. The best NBA systems treat the customer's life as the main storyline and the bank's products as supporting characters.

Under the hood, recommendation engines come from the same family tree as Netflix and Amazon, but they grow differently in a regulated environment. The classic approaches are straightforward to describe in plain language:

Content-based recommendations look at the attributes of a product or action and match them to what the customer has shown they like or need. If a customer keeps overdrafting, a content-based system might recommend an overdraft buffer feature or a small-dollar credit line (where appropriate), because those products are explicitly designed to address that pattern.

Collaborative filtering is the "people like you also liked" approach. It doesn't need to understand why something works; it observes that certain customers behave similarly and that certain actions tend to land well with that cluster. In banking, this can be powerful and dangerous at the same time, because "people like you" can unintentionally become a proxy for protected characteristics. You can build a recommendation engine that is statistically clever and socially reckless without ever mentioning race, gender, or age.

Hybrid systems blend the two: they use product attributes, customer behavior, and population-level patterns together, often with business rules layered on top. In practice, almost every bank-grade recommender ends up hybrid because pure approaches break on real-world constraints: compliance rules, product eligibility, channel constraints, and the simple fact that banks can't recommend what the customer can't get.

That last line—*can't recommend what the customer can't get*—is not just a technical constraint. It's a trust constraint. A customer who is offered a card they don't qualify for learns a lesson quickly: the bank isn't paying attention, and the "personalisation" is theater. Eligibility logic and policy rules are the unglamorous backbone of relevance. The model can score the appeal of an action, but a rules layer should filter for what is permissible, suitable, and operationally executable. Think of it as the same philosophy that makes a conversational bot escalate rather than hallucinate: if the system can't do the thing safely, it shouldn't tease it.

Banks also have a special category of "recommendation" that retailers don't: advice-adjacent nudges. A streaming service can be wrong about a movie and nobody files a complaint. A bank that nudges a customer to move money, change repayment behavior, or open credit can unintentionally cross into the territory of financial advice, or at least be perceived as giving it. That perception matters. If a model-driven nudge leads a customer into fees or hardship, the customer won't care whether the bank called it "guidance" rather than "advice." They'll care that the bank pushed them.

So the safest, most valuable starting point for personalisation is often not product offers. It's *friction removal*. When the model sees that a customer is repeatedly making manual payments, it can suggest autopay. When it sees a recurring subscription that has doubled in price, it can flag it. When it sees a pattern of late fees, it can suggest due-date changes or reminders. These are not glamorous. They are, however, the kinds of interventions that feel like the bank is finally on the customer's side.

Personalisation also thrives when it's framed around goals rather than offers. "Save more" is a scold. "You're $42 short of your 'Emergency Fund' target this month; want to move it now?" is a concrete option. This is where banks can borrow from behavioral economics without becoming manipulative. People respond better to specific, timely prompts than to generic motivation posters. The trick is to keep the system respectful: a nudge should feel like a door opening, not a hand pushing.

A subtle design choice separates helpful nudges from annoying ones: *the option to dismiss and the willingness to remember.* If a customer dismisses a travel-insurance prompt three times, the model should learn that this is not their thing. If a customer ignores every overdraft message, the system should change strategy: perhaps offer a different solution, or stop interrupting. The fastest way to make personalisation feel creepy is to behave like you aren't listening. The fastest way to make it feel broken is to behave like you have amnesia.

Getting this right requires treating the recommendation engine as a real production system with governance, not as a marketing widget. Banks already know how to govern models when money is on the line: credit scoring, fraud detection, liquidity forecasting. Personalisation should sit in the same mental bucket because it can cause harm in quieter ways: pushing vulnerable customers toward expensive products, creating disparate outcomes across groups, or generating complaint spikes because a model learned the wrong lesson from historical behavior.

That's why model risk management ideas belong here, even if the outputs look like "soft" suggestions. A recommendation engine has assumptions, blind spots, and failure modes. It can drift when customer behavior changes (think of how spending patterns shift during economic shocks). It can be misused when a business team changes a campaign goal without revisiting the model's constraints.

It can be misunderstood when frontline teams interpret a model score as certainty. A mature bank treats those risks as first-class, with documentation, monitoring, and independent challenge.

Fairness is the hardest guardrail to implement because it doesn't come with a single definition. In lending, regulators and risk teams have well-developed frameworks to think about discrimination, disparate impact, and protected classes. In personalisation, the line is fuzzier but the stakes remain. A model that offers fee waivers to customers who complain the loudest may disadvantage those who are less comfortable escalating. A model that recommends credit-building products only to customers who already use credit frequently may widen inclusion gaps. A model that learns from historical acceptance rates may keep offering premium products to the same populations that have always been targeted, even if eligibility has broadened.

The uncomfortable truth is that "optimize for conversion" is a fairness strategy only if you believe the past was fair. Banks can't afford that assumption. Guardrails here look like: limiting the use of sensitive or proxy variables, testing outcomes across segments, constraining what products can be recommended to whom, and monitoring the downstream effects (fees paid, complaints, retention) rather than only clicks. Personalisation isn't a game of higher open rates; it's a long game of trust.

Privacy is the other guardrail that can't be bolted on at the end. The raw material for personalisation is intimate: income timing, medical payments, family spending patterns, location hints embedded in merchant data. Customers may have consented in the fine print, but consent is not the same as comfort. A good rule of thumb is *contextual integrity*: only use data in ways that match the customer's reasonable expectations in that moment. If the customer is in the travel section of the app, a travel-related nudge feels coherent. If the customer is trying to deposit a check and gets an unrelated prompt about debt consolidation, the bank feels intrusive.

Even when the nudge is appropriate, the *wording* is a privacy choice. "We noticed you were at the airport" is a different sentence than "Planning a trip?" The first is surveillance-flavored. The second is respectful. The model may have used the same underlying signals, but the customer experiences them differently. Banks should treat prompt copy as part of the control environment: reviewed, tested, and tuned to avoid revealing sensitive inference.

The smartest banks also accept a paradox: personalisation gets better with data, but trust gets worse when customers feel over-modeled. That's why a lot of high-performing recommendation systems in banking deliberately use *coarse* features when they can. You don't always need a precise merchant category breakdown if a simple "travel spend increased" signal will do. You don't always need to infer life events if the customer explicitly tapped 'Buying a home' in the app. Explicit preferences are gold: they're accurate, and they feel consensual.

Operationally, the recommendation engine is usually a pipeline with three layers. The first layer is signal generation: turning raw transactions and events into features (paycheck detected, subscription increased, savings balance trending down, international spend spike). The second layer is scoring and ranking: predicting which action will be helpful or accepted. The third layer is orchestration: deciding *where* and *when* to deliver the message (push notification, email, in-app card, agent suggestion), and enforcing frequency caps so the bank doesn't become background noise.

Orchestration is where a lot of "AI" projects go to die, because it's not glamorous and it's full of politics. Marketing wants more impressions. Service teams want fewer distractions. Risk wants tighter constraints. Product teams want messages to appear inside the app experience they own. Customers want the bank to stop spamming them. The recommendation engine sits in the middle like a traffic cop, and the only way it survives is with clear priorities: customer benefit first, compliance always, business value as a consequence rather than the only goal.

A surprisingly effective pattern is to treat NBA decisions like a newsroom treats headlines: there is limited real estate and the bar is high. The app home screen can't hold twenty offers without becoming a cluttered flea market. A push notification that isn't urgent is a small act of disrespect. The discipline of scarcity improves quality. It forces the model to choose, and it forces the business to accept that "do nothing" is sometimes the best action. In a world where every team wants a slot, the model's restraint becomes a competitive advantage.

When personalisation is deployed in contact centers, the dynamics get even more interesting. An agent on a call is not going to read a long recommendation rationale. They need a short suggestion that fits the moment: "Customer is traveling; remind about foreign transaction fee-free card" or "Customer called about late fee; offer due-date change option." This is NBA as agent-assist: not a direct pitch, but a service prompt that helps the human resolve the issue and maybe deepen the relationship in a way that feels natural. If the suggestion is tone-deaf, the agent will ignore it and quietly lose faith in the system. If the suggestion is consistently helpful, it becomes part of how the bank does service.

The most dangerous failure mode in personalisation is the one that looks like success on a dashboard. A model can improve click-through rates by becoming more sensational, more frequent, or more financially aggressive. The dashboard cheers. The complaint inbox fills. The bank doesn't connect the dots until churn shows up months later. That's why evaluation has to go beyond offline metrics and immediate conversions. The bank should watch for second-order signals: repeat contact rates, complaint rates, opt-outs from messaging, fee incidence, and long-term retention. If the model is truly creating value, it should reduce friction and increase loyalty, not just harvest attention.

Testing matters too, and banks should borrow the scientific mindset from digital businesses: controlled rollouts, A/B tests where feasible, and staggered deployment when perfect randomization is hard. Without that discipline, it's easy to attribute every improvement to the shiny new model when the real driver was a product fix, a rate change, or a seasonal spending pattern. Personalisation lives in a noisy world; good measurement is the difference between learning and storytelling.

There's also a human factor that models can't replace: dignity. Customers aren't "users" in the abstract; they're people with rent due, kids' tuition, medical bills, and the occasional impulsive splurge that makes life bearable. A recommendation engine that treats every liquidity dip as an upsell opportunity will feel predatory. A system that recognizes financial stress and shifts into a supportive posture—offering budgeting tools, payment flexibility options, or a gentle nudge to talk to a human—can make a bank feel like an institution worth staying with. That posture isn't just good ethics. It's good business, because trust is the only sustainable moat in a category where products are easily copied.

The real promise of personalisation in banking is not that the bank will know the customer perfectly. That fantasy ends in creepiness or compliance trouble. The promise is smaller, more practical, and more valuable: fewer irrelevant interruptions, more timely help, and a sense that the bank is paying attention when it matters. The technology that makes that possible is sophisticated, but the outcome is simple. When the recommendation is right, it doesn't feel like marketing. It feels like competence.

5.3 Voice and Text Analytics

Banks are sitting on one of the world's richest collections of customer feedback, and for decades they treated it like nuclear waste: store it somewhere safe, touch it as little as possible, and hope nobody asks questions. Not because the feedback wasn't valuable, but because it arrived in the least spreadsheet-friendly format imaginable. Phone calls. Free-text emails. Chat transcripts full of typos, sarcasm, half-finished thoughts, and the occasional all-caps meltdown. A bank could measure what it could count—call volume, average handle time, abandonment rate—but the content itself stayed mostly dark. The customer was talking; the institution was listening only to the dial tone.

Natural Language Processing (NLP) flips on the lights. NLP is the set of techniques that let computers work with human language: turning audio into text, extracting meaning, spotting themes, and measuring sentiment (roughly: how positive or negative the language sounds). The reason this matters for customer experience is simple: surveys are a postcard from a tiny minority of customers, often the happiest or angriest. The real voice of the customer is the 2:13 a.m. chat where someone says, *My paycheck is missing and rent is due tomorrow.* That sentence is not a data point until you make it one.

The first step is transcription: converting speech to text. Once calls become text, they become searchable, analyzable, and—with the right controls—measurable at scale. That's the industrial breakthrough. A supervisor can't listen to 10,000 calls a day, but a model can read 10,000 transcripts before lunch and report back: *Password reset failures up 28% since 9 a.m.* The contact center stops being a place where problems are discovered by rumor ("Are you hearing more complaints about the app today?") and becomes an early warning system with timestamps.

The emotional shift inside a bank is surprisingly dramatic. Before analytics, a product team might spend weeks debating whether an app update "feels buggy." After analytics, the debate becomes short and specific: *Complaints mentioning "deposit" and "camera" spiked within 30 minutes of release 6.4.1 on iOS.* The institution moves from vibes to evidence. And because money problems are time-sensitive, evidence that arrives two days earlier is not a rounding error. It can be the difference between a contained incident and a weekend-long reputational fire.

Voice and text analytics generally fall into four buckets, each with its own strengths and traps: intent detection, entity extraction, sentiment/emotion signals, and summarization. Banks rarely deploy them all at once. They start with the one that maps to an obvious operational pain point—and then they discover that language is slippery, and every metric needs a grown-up in the room.

Intent detection answers: what is the customer trying to do? "Dispute a charge," "reset password," "close account," "stop payment." At scale, intent detection is the foundation of routing. It decides whether a customer ends up with the right queue, the wrong queue, or the purgatory queue where they repeat themselves three times. A good intent model reduces transfers (the customer's least favorite sport). A bad one creates them. The trick is that customers don't state intent in neat labels. They tell stories. They start in the middle. They change direction. "I was trying to pay my credit card and it said error and now I'm worried it'll be late" is at least three intents and one emotion, all braided together.

Entity extraction answers: what are the key details? Amounts, dates, merchant names, confirmation numbers, account types, locations. Entities are the difference between "customer called about fraud" and "customer reported $812.43 at Merchant X on 2026-02-28; card present; travel context." When entity extraction is reliable, it becomes the scaffolding for automation: prefilling dispute forms, generating case notes, and giving agents a head start. When it's unreliable, it becomes a quiet source of errors—a misplaced decimal point is the kind of bug that turns a process improvement into a complaint.

Sentiment and emotion signals attempt to quantify how the interaction feels. Done well, they let a bank detect frustration spikes early, trigger supervisor support in real time, and identify policies that consistently cause anger. Done poorly, they turn into horoscope numbers that get pasted into dashboards and ignored. Sentiment is not truth; it's a proxy. Customers can be polite and furious. They can be loud and joking. Some cultures express distress differently. A sentiment model that isn't validated across real customer populations will inevitably misunderstand someone—and the bank will make operational decisions based on that misunderstanding.

Summarization is the new darling: an AI that reads a long call transcript and writes the case notes. This can be transformational for operations because after-call work is a tax on every interaction. Agents often spend precious minutes writing what happened, what was promised, and what needs follow-up. A good summary frees time for more conversations or simply reduces burnout. But summarization introduces a new class of risk: the summary becomes the record. If it's wrong, it's *officially* wrong. The bank needs controls to make sure summaries are clearly labeled as AI-generated, easy to edit, and sampled for accuracy. In regulated workflows, the summary can't be a confident storyteller; it has to be a faithful stenographer.

The most powerful part of analytics isn't any single model output. It's what happens when you connect them into a feedback loop. A bank can detect a spike in the intent "can't log in," observe that sentiment is sharply negative, extract the entity "error code 403," and find that the theme "after updating" appears in the same cluster. Suddenly the bank has a crisp narrative: a new app release broke login for a subset of customers, and the contact center is feeling the impact in real time. That narrative is operational gold. It gives product teams something to fix and comms teams something honest to say.

Honesty matters because customers can smell corporate fog. When a bank's status page says "We are aware some customers may be experiencing intermittent issues," customers hear, *We are hiding how bad it is.* Analytics can help banks speak plainly: "Some customers on Android are seeing login failures after version 6.4.1. We're rolling back and expect resolution by 3 p.m. ET." The fastest way to reduce negative sentiment is not to train the chatbot to apologize better. It's to fix the problem and communicate like a competent adult.

There's a second use case that doesn't get as much attention: quality assurance that isn't performative. Traditional QA in contact centers is often a thin layer of monitoring: sample a few calls, score agents on a checklist, file it away. Analytics can turn QA into something closer to continuous improvement. If models can identify calls where the agent interrupted repeatedly, failed to mention a required disclosure, or misclassified a dispute, the bank can coach with precision rather than generalities. That's good for customers and often good for agents, who are tired of being judged by blunt averages.

The danger is turning analytics into a surveillance state. If every sentence an agent says becomes a metric, the job becomes acting. Agents start optimizing for the score rather than for the customer. They avoid creative problem-solving because it might sound "off script" to the model. Banks need to be explicit about purpose: analytics should be used to support agents and improve systems, not to punish individuals for being human in messy conversations. The fastest way to sabotage adoption is to deploy a tool branded as "agent assist" that is experienced as "agent police."

Sensitive data redaction is the enabling control that makes the whole analytics machine feasible. Calls and chats inevitably contain personal information: Social Security numbers, card numbers, addresses, dates of birth. If transcripts are stored raw and widely accessible, the bank has created a new risk surface: an internal dataset of highly sensitive information that can be misused or breached. Redaction tools attempt to remove or mask that data before it spreads through analytics pipelines and training sets. This isn't just compliance housekeeping. It's the difference between a scalable program and a pilot that never leaves the lab because nobody wants to sign off on the risk.

Redaction also forces a trade-off that banks must handle thoughtfully: the more you mask, the less useful some analytics become. If you redact all numbers, you might lose the ability to detect patterns like "overdraft fee $35" or "chargeback $812.43." Good implementations are surgical: redact what's truly identifying (full card numbers, SSNs), preserve what's operationally meaningful (rounded amounts, general categories), and keep the keys to the kingdom tightly controlled. Privacy-by-design is not a slogan here; it's architecture.

Once transcripts are safe enough to analyze, banks can build what amounts to a *theme radar*. Theme detection clusters conversations into recurring topics, even when customers use different words. "App keeps crashing," "screen goes blank," and "it closes when I deposit a check" can be grouped into a single theme: mobile crash during deposit flow. This is where NLP earns its keep as a listening device at scale. It doesn't just count complaints; it finds the shape of them.

The strategic payoff is that a bank can see around corners. A small rise in a new theme can foreshadow a larger incident. A subtle increase in hardship-related language can hint at economic stress in a customer segment. A spike in "I never received the OTP" can indicate a vendor issue with SMS delivery. The contact center becomes a kind of seismograph for operational and financial stress, and the bank gets the chance to respond before the tremor becomes an earthquake.

That said, analytics outputs are still model outputs, which means they deserve the same skepticism banks apply elsewhere. A sentiment score is not a fact; it's an estimate produced by a model trained on some dataset that may or may not resemble the bank's real customers. An intent classifier can drift when product names change or when scammers adopt new scripts. Theme models can over-cluster, under-cluster, or miss emerging issues entirely. Treating these systems as "just analytics" is how banks end up with dashboards that look sophisticated and decisions that are naïve.

A mature bank validates language models like it validates credit models: test performance, monitor drift, and document limitations. Transparency matters, too. If a supervisor is alerted because a model flagged a call as "high-risk," the supervisor should understand what triggered it: keywords, acoustic cues, conversation patterns. Otherwise the organization starts making decisions based on black-box vibes, which is exactly what analytics was supposed to replace.

There's also a practical truth that gets missed in vendor demos: the model is only as good as the plumbing. If the bank's call recordings are incomplete, if channel data isn't linked, if timestamps are wrong, if customer IDs don't match across systems, then even the best NLP becomes an expensive microscope pointed at a blurry slide. Data governance—the boring discipline of making sure data is accurate, consistent, and traceable—is what makes voice-of-customer programs trustworthy. Without it, the bank builds an insights engine that confidently explains yesterday's problems while today's customers wait on hold.

One of the most valuable artifacts produced by voice and text analytics is the *conversation summary* that travels with the customer. Not a fluffy recap, but a structured note that answers: what happened, what was decided, what was promised, and what the next step is. When summaries are done right, they reduce repeat contacts because customers don't have to retell their story and agents don't have to reconstruct it from scraps. They also create accountability inside the bank. A promise made in a call becomes visible to the next agent. The institution starts behaving like one continuous service experience rather than a series of disconnected encounters.

Summaries also expose a cost that banks sometimes ignore: AI isn't free once you turn it on. Transcription costs money. Real-time analytics costs money. Summarization costs money. At small volumes, it's negligible. At millions of interactions, it becomes a line item worth modeling. The irony is that the very features that improve customer experience—the ones that reduce agent effort and improve handoffs—are the ones that rack up usage fees. That doesn't make them bad ideas. It just means finance teams need to treat AI like any other operational capability: estimate unit costs, tie them to benefits, and avoid the trap of "We saved labor" while quietly adding a new variable cost that grows with every interaction.

The best voice and text analytics programs have a distinctive rhythm. They don't just produce reports. They produce *actions*. A theme spike triggers an incident review. A persistent complaint cluster triggers a product fix. A confusing policy explanation triggers a rewrite in plain language. A pattern of dispute mishandling triggers training and workflow redesign. Analytics becomes the connective tissue between customers' messy words and the bank's structured change processes.

And that's the real reason NLP matters for growth. Growth in banking is often framed as acquiring more customers or selling more products. But retention is growth too, and retention is mostly about avoiding preventable pain. When customers leave a bank, they rarely announce, "I am churn caused by a 1.8-point drop in satisfaction." They leave because something broke, nobody fixed it quickly, and the institution felt indifferent. Voice and text analytics help a bank catch indifference early, when it still looks like a small spike in angry sentences rather than a permanent loss of trust.

Language is where banking becomes personal: where "funds availability" becomes "I can't buy groceries," where "authentication failure" becomes "I'm locked out of my own money," where "dispute intake" becomes "someone drained my account." NLP doesn't replace the human part of that. It makes sure the bank can hear it, quantify it, and respond at the speed modern customers expect—without drowning in its own conversations.

5.4 Value Measurement and ROI

The most dangerous number in customer-facing AI is also the easiest to calculate: "calls deflected." It looks clean on a dashboard and comforting in a budget meeting. If the bot answered 30% of inbound chats without a human, the spreadsheet declares victory. But deflection can be a magic trick: the call didn't disappear, it just got pushed somewhere you're not measuring. It might come back as a repeat contact tomorrow. It might show up as a branch visit (expensive). Or it might vanish into the competitor's app-no ticket, no transcript, no opportunity to recover.

Banks have been here before. Every generation of "efficiency" tools-from IVR trees to offshore scripts-has produced the same temptation: optimise what's easy to count and ignore what customers actually feel. Customer-facing AI raises the stakes because it can *sound* competent while quietly creating new failure modes: confidently wrong answers, compliance landmines, and friction that feels personal ("Why doesn't my bank understand me?"). The only antidote is measurement that treats AI like a product launch *and* an operational change *and* a risk event, all at once.

ROI is not one number; it is an argument

Return on investment isn't a metric; it's a story you can defend under cross-examination. The CFO will ask, "Did it save money?" The head of service will ask, "Did it make the queue shorter without making customers angrier?" Compliance will ask, "Did it create new ways to break rules?" And a frontline manager will ask the question that actually decides whether the initiative survives: "Did it make my agents' day better or worse?"

A sturdy ROI argument has three pillars:

1. *Efficiency*: What changed in cost-to-serve and capacity?

2. *Quality*: What changed in outcomes customers care about-resolution, effort, trust?
3. *Growth and risk*: What changed in retention, share of wallet, complaints, and regulatory exposure?

If any pillar is missing, the argument tips over. Efficiency alone invites a race to the bottom ("We can deflect more!"). Quality alone risks being inspirational but unfunded. Growth without risk becomes the kind of optimism regulators love to correct.

The trap of "cheaper conversations"

Banks often begin with a simple equation:

$$\text{Savings} \approx (\text{Contacts Deflected}) \times (\text{Cost per Contact})$$

It's not wrong. It's just incomplete in the same way "calories in, calories out" is incomplete advice for nutrition: technically true, operationally misleading.

Three reasons that equation exaggerates ROI:

- *Not every deflected contact removes a unit of labour.* If staffing is fixed by scheduling constraints, contracts, or minimum coverage, the "saved" minutes might not translate into fewer paid hours for months.
- *Deflection can boomerang.* A customer who fails in self-service may return twice-once to try again, then again when they're angry. The contact volume doesn't fall; it reshapes.
- *AI has operating costs.* Tokens, per-interaction fees, vendor licences, model monitoring, red-teaming, and prompt/content maintenance aren't rounding errors at scale. A bot that "costs pennies per chat" can still become a material line item when you multiply it by millions of chats and layer on governance.

A more honest view treats conversational AI like an additional channel with its own cost and failure rates, not a magical vacuum that sucks demand out of the call centre.

Start with a service economics map, not a dashboard

Before choosing KPIs, map the service system the way an engineer would: inputs, queues, constraints, and bottlenecks. Contact centres are factories for problem-solving, and their output is not "calls handled." The output is "customer issues resolved with acceptable risk."

A practical service economics map answers questions like:

- Which contact types are *simple* (address change) versus *high-consequence* (fraud claim, mortgage hardship)?
- Where does time go-talk, hold, after-call work, follow-ups?
- Where does failure happen-handoffs between teams, authentication, broken app flows?
- Which interactions create revenue opportunities (card upgrade, retention save) versus purely defensive cost?

Once the system is mapped, the KPI set becomes obvious: measure the bottlenecks, not the noise.

A balanced scorecard that executives actually use

A balanced scorecard is just a promise to measure more than one thing, so you don't "win" by breaking the business. The trick is to keep it tight enough that leaders will look at it weekly.

Efficiency metrics (capacity and cost). These are the numbers that keep the lights on:

- *Average Handle Time* (*AHT*): total agent time per contact, usually talk + hold + wrap/after-call work.

- *Containment/automation rate*: share of interactions resolved end-to-end without an agent (define "resolved" carefully).
- *Agent utilisation and occupancy*: are agents busy in the right way, or drowning in peak loads?
- *Cost per resolved issue*: the unit that matters more than cost per contact.

AHT is seductive because it moves quickly. But chasing AHT can produce perverse incentives: agents rush, customers call back, and the bank pays twice.

Quality metrics (did the customer get what they came for?). If AI makes service cheaper but less reliable, customers don't award you a trophy-they leave.

- *First Contact Resolution (FCR)*: proportion of issues resolved without repeat contact in a defined window.
- *Repeat contact volume*: the shadow cost of failure.
- *Transfer/escalation rates*: how often AI hands off to humans, and how often humans hand off to other humans.
- *Customer Effort Score (CES)* or similar: how hard did the customer have to work?

FCR deserves special respect and special skepticism. Respect, because it aligns with the customer's goal ("solve it once"). Skepticism, because definitions vary: is it "no repeat contact within 7 days"? Does a new channel count as repeat? If a customer calls back for a related issue, is that a failure? Without a consistent definition, FCR becomes a number you can argue about forever and improve by changing the rules.

Trust and risk metrics (are we creating new liabilities?). Customer-facing AI adds a new risk surface: it speaks. That means it can mislead, discriminate, or disclose.

- *Compliance exception rate*: instances where the system produced disallowed content or omitted required disclosures.
- *Complaint rate and severity*: especially complaints tied to misinformation, tone, or accessibility barriers.
- *Fraud and social engineering signals*: whether the channel is being exploited (or unintentionally helping attackers).
- *Privacy leakage incidents*: including accidental exposure in summaries, transcripts, or agent-assist side panels.

These metrics are often sparse-thankfully. But sparse doesn't mean ignorable. A single high-severity incident can wipe out a year of efficiency gains.

Growth metrics (the only ROI that compounds). Growth metrics are where AI can become more than a cost play. They're also where self-deception is easiest, because many factors move at once.

- *Retention and churn*: do customers who use the AI-assisted journey stay longer?
- *Cross-sell/upsell conversion*: did better service increase product adoption, and did it do so fairly?
- *Share of wallet proxies*: direct deposit stickiness, card spend, savings balances-used carefully.
- *Organic referral signals*: complaints going down is good; unsolicited praise and referrals are better.

Net Promoter Score (NPS) often sits here, like a familiar flag on the hill. Many executives like it because it compresses sentiment into a single number and, in some studies and consulting narratives, correlates with growth within competitive sets. Critics counter that the relationship is inconsistent, context-dependent, and not a causal proof. Treat NPS the way you treat a credit score: useful as a signal, dangerous as a sole decision-maker.

Define "resolved" like you mean it

If a bot tells a customer, "Your dispute has been filed," but the case never opens correctly, you don't have containment-you have delayed failure. Measuring AI requires definitions that match real outcomes:

- *Resolution* should mean the underlying task completed in core systems (a payment stopped, a card replaced, a fee waived under policy), not merely "the customer stopped typing."
- *Successful handoff* should mean the human agent receives context, the customer does not repeat themselves, and the issue is handled faster or better because the AI did pre-work.
- *Appropriate deflection* should mean the customer chose self-service because it worked, not because the bank hid the phone number behind five screens.

These definitions feel picky until the first time the bank celebrates a "40% containment rate" while complaints spike and agents mutter that the bot is "a liar with confidence."

Causality: the difference between proof and vibes

A common measurement mistake goes like this: the bank launches AI in March, and by May the call centre is cheaper. Victory! Except May also had lower fraud volume, a new staffing model, and a mobile app redesign that reduced password resets. The AI might deserve credit-or might be riding in the back seat while someone else drives.

Causal measurement is simply the discipline of isolating what changed because of the AI, not because of the weather.

One practical approach is a *staggered deployment*. Roll out agent-assist or a new bot capability to a subset of teams, branches, or regions first, leaving others as a comparison group for a defined period. Measure productivity outcomes (issues resolved per hour, AHT, after-call work time) alongside quality outcomes (FCR, customer sentiment, complaint rates). If the early groups improve while the comparison group stays flat-and the pattern persists as the rollout expands-you have something closer to evidence than enthusiasm.

This approach has two virtues that executives appreciate:

- It is operationally realistic. Banks already roll out changes in waves.
- It creates a narrative you can defend: "Here is what happened where we turned it on, relative to where we didn't."

It also forces uncomfortable clarity. If the numbers don't improve in the treatment group, there's nowhere to hide. That's a feature, not a bug.

The cost model banks forget to finish

Traditional contact centre ROI models focus on labour, because labour is big and visible. AI adds a new category: variable compute and vendor costs tied to usage, plus fixed costs tied to governance. A complete cost model typically includes:

- *Variable AI costs*: per-interaction LLM fees, speech-to-text, text-to-speech, summarisation, translation, retrieval, and any third-party tools invoked.
- *Fixed platform costs*: licences, integration work, observability, secure logging, evaluation tooling.
- *People costs*: prompt/content owners, conversation designers, model risk management, compliance review, red-team testing, and the engineers who keep the pipes from leaking.

- *Failure costs*: rework from wrong answers, increased handle time when the AI confuses the customer, goodwill credits or fee reversals, and potential remediation after incidents.

A useful mental model is to treat AI like a junior employee who works at the speed of software: cheap per minute, but requiring supervision, training materials, and guardrails. The "salary" is tokens; the management overhead is real.

A practical ROI table for decision-making

When leadership asks "Is it worth it?", they're really asking for a comparison between two futures: the service operation with AI and the service operation without it. Putting that into a compact table makes the argument concrete.

Category	*What to measure*	*What can go wrong if mismeasured*
Efficiency	AHT, cost per resolved issue, automation rate, agent capacity	You optimise speed, create repeat contacts, and call it "savings"
Quality	FCR, repeat contact within window, transfer rate, customer effort	You celebrate containment while customers quietly fail and come back angrier
Trust/Risk	Compliance exceptions, complaint severity, privacy incidents	You ship a persuasive system that is occasionally illegal
Growth	Retention, card spend, product adoption, referral signals, NPS (as a signal)	You attribute broad market movement to AI and over-invest in the wrong lever

Table 5.1: A balanced measurement view of customer-facing AI ROI.

The table is simple on purpose. The sophistication comes from instrumentation and definitions, not from fancy math.

Quality-adjusted efficiency: the metric executives deserve

If one number must go on a slide, make it harder to game. A "quality-adjusted cost to serve" metric punishes cheap service that fails.

One example:

$$\text{Quality-Adjusted Cost} = \frac{\text{Total Service Cost}}{\text{Resolved Issues}} \times (1 + \alpha \cdot \text{Repeat Rate})$$

Where α reflects how painful repeat contacts are for the business (and for customers). The exact formula matters less than the philosophy: reward resolution, penalise rework. It nudges teams away from shallow wins.

Measuring the handoff, not just the bot

Many banks treat human handoff as an operational afterthought: if the AI can't answer, transfer to an agent. But handoff is where customer trust is either rescued or destroyed.

Measure handoff like a first-class product:

- *Handoff latency*: how long until a human is engaged after escalation triggers?
- *Context integrity*: did the agent receive the summary, identity signals, and customer intent accurately?
- *Customer repetition rate*: how often customers restate their problem after transfer (a direct proxy for a broken baton pass).
- *Post-handoff AHT and FCR*: if AI pre-work is real, these should improve.

A surprising pattern often emerges: the best ROI does not come from eliminating humans; it comes from making humans dangerous in the good way-faster, more consistent, less burdened by searching and summarising.

Sentiment is an early-warning system, not a trophy

Banks love lagging indicators because they feel definitive: quarterly churn, annual complaint totals, monthly cost per contact. The problem is that lagging indicators are where you learn you were wrong *after* the damage spreads.

Leading indicators-especially sentiment from calls and chats-act like smoke detectors. They don't tell you the full story, but they tell you where to look *now*. A spike in negative sentiment after a bot update can reveal a broken flow within hours. If the bank waits for monthly reports, the story will already be circulating on social media.

The measurement discipline here is to connect sentiment to action:

- Identify the top emerging complaint themes weekly.
- Tie each theme to an owner (product, operations, policy).
- Track "time to fix" as a metric in its own right.

AI that listens without prompting the organisation to respond is just surveillance with better charts.

Fairness and inclusion: ROI's quiet constraint

A bank can accidentally build an ROI engine that works by excluding hard customers. If the AI experience is smooth for digitally fluent customers but brittle for customers with limited English proficiency, disabilities, or low digital trust, the cost curve might look great-until regulators, advocates, or the market forces the bank to confront who was left behind.

Measurement needs segmentation, not as a marketing flourish but as a safety rail:

- Track resolution, escalation, and sentiment by language, channel, and customer segment.

- Monitor whether AI steers certain customers toward slower or more restrictive paths.
- Audit outcomes for protected-class proxies where legally and ethically appropriate, using the same seriousness applied in lending models.

Inclusion isn't separate from ROI. It *is* ROI, because reputational damage, remediation costs, and lost lifetime value are expensive forms of "efficiency."

What "good ROI" actually looks like in practice

Good ROI rarely looks like a dramatic single metric. It looks like a cluster of improvements that reinforce each other:

- Automation rises *modestly* in high-confidence, low-risk intents.
- FCR rises because the easy stuff is genuinely solved and the hard stuff is routed better.
- Agents handle fewer "where is my payment?" calls and more complex cases, with better tools and lower burnout.
- Complaints about misinformation fall because guardrails improve faster than capabilities expand.
- Retention and product adoption lift slightly, then compound over time.

Bad ROI, by contrast, has a distinct smell: containment up, transfers messy, repeat contacts up, agents frustrated, and a growing gap between the dashboard and what customers say when they finally reach a human.

A final discipline: pre-mortems and kill switches

Every AI programme should begin with a pre-mortem: imagine the headline of the failure and work backwards. "Bank chatbot told customers to share passwords." "Bot misquoted hardship eligibility." "AI escalations doubled and broke the queue." These are not paranoid fantasies; they're predictable failure modes of systems that speak in fluent sentences.

Measurement should include explicit thresholds that trigger action:

- If compliance exceptions exceed X per 10,000 interactions, roll back the new capability.
- If repeat contacts rise Y% for intents handled by the bot, investigate and constrain those intents.
- If sentiment drops sharply after a release, pause expansion until root cause is found.

The most persuasive ROI story is not "we launched AI and hoped." It's "we launched AI with instruments, guardrails, and the humility to shut it off when it misbehaved." In banking, that humility isn't just good governance-it's a competitive advantage, because trust is the one asset you can't refinance after you spend it.

Chapter 6

Governing and Deploying AI Confidently

If a traditional statistical model is a reliable sedan, a machine learning model is a Formula 1 car: faster and more powerful, but disastrous if you lose control on a sharp turn. This chapter moves beyond the code to the critical infrastructure-governance, operations, and compliance-that keeps the bank from crashing. We explore how to validate 'black box' decisions, handle the shifting sands of global regulation like the EU AI Act, and build an operational heartbeat (MLOps) that ensures your models don't degrade the moment they hit production.

6.1 Building the Guardrails: Model Risk Management

Bankers love certainty. The contract says the borrower pays *this* amount on *that* date, and if they don't, the bank has remedies. Model risk management (MRM) exists because the real world refuses to be that tidy. Now add machine learning: a class of models that are often *probabilistic* by design, trained on yesterday's patterns, and quietly sensitive to tiny shifts in data. The result is a new kind of fragility: not the obvious "the code is broken" failure, but the more dangerous "the code is working perfectly and producing the wrong confidence" failure.

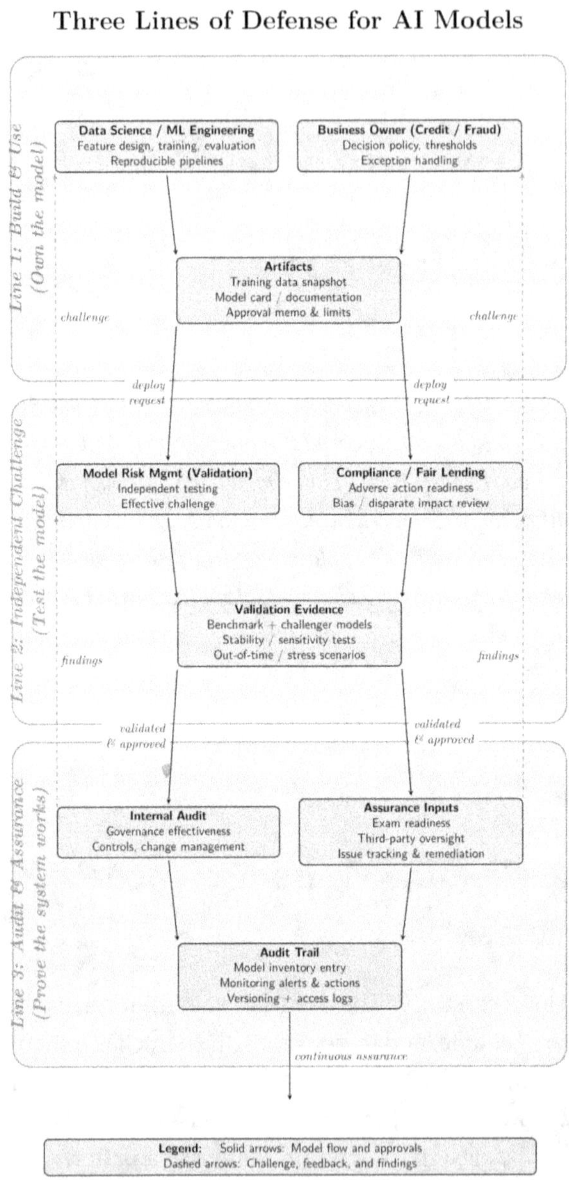

The "three lines of defense" is an old idea with a stubbornly modern punchline: the people who *build and run* a system can't be the only people who *judge* it. In a pre-ML world, the first line might have been a credit policy team and a scorecard developer; the second line would validate the model; the third line would check that everyone followed the playbook. With AI, the cast expands, the props multiply, and the plot gets twistier—but the logic holds. The trick is making the lines strong enough to stop bad outcomes without turning innovation into a paperweight.

MRM is easiest to understand by admitting what makes AI different. Many machine learning models are not "wrong" in the way a broken calculator is wrong. They're wrong in the way a confident weather forecast can be wrong: they were trained on patterns that used to hold, and reality changed its mind. A model can be statistically elegant and operationally reckless, especially when it's dropped into a lending workflow where a few basis points of performance look small on a dashboard and enormous in a lawsuit.

The probabilistic nature of ML creates a particular failure mode: *the model's confidence becomes a product feature*. A logistic regression scorecard will give you a probability too, of course, but ML systems—especially ensembles and neural networks—often feel more "alive" to their users. They update, they learn, they find signal in scraps. That's exactly why banks need guardrails that treat probability as a volatile material. You don't store gasoline in a paper cup just because it's useful.

A good MRM framework starts with a deceptively basic question: *what counts as a model*? In many banks, the answer has expanded from "the PD model" to "anything that turns inputs into an estimate used to make a decision." That includes credit decision engines, fraud classifiers, collections prioritization, marketing propensity models, and sometimes even "helpful" AI systems that summarize customer interactions for a banker who is about to recommend a product. If the output changes behavior, it creates risk—even if no one wants to call it a model because the word triggers governance.

The next question is less philosophical and more practical: *how bad would it be if this model were wrong*? Classic MRM loves tiering: low-risk models get lighter validation; high-risk models get deeper testing, tighter change control, and louder oversight. AI makes the tiering conversation sharper because the blast radius can grow without a corresponding increase in headcount. Automate one decision, and you scale the decision; automate *the wrong* decision, and you scale harm.

A surprisingly common failure in AI programs is not statistical. It's clerical. Teams can't reliably answer: "Which version is in production, what data was it trained on, and who approved the last change?" That is not a question about mathematics. It's a question about discipline. A model inventory—maintained like a living system, not a spreadsheet graveyard—is the first guardrail that actually earns the name.

Documentation is the next guardrail, and it should be treated like a deposit agreement: boring until it saves you. The best AI documentation doesn't try to prove the model is brilliant. It tries to make the model *legible* to someone who didn't build it and is paid to be skeptical. For ML, "legible" means at least five things: what the model is for (and not for), what data it uses (and excludes), how it was trained and evaluated, what its known weaknesses are, and what monitoring will catch those weaknesses when they get worse.

Traditional validation methods often fail because they assume the model is a stable object. Neural networks and complex ensembles can be stable in aggregate yet twitchy in the corners. Change a preprocessing step, reorder categories, adjust an imputation rule, and performance might barely move—while fairness metrics or adverse action explanations become incoherent. In lending, that's not a technical nuisance; it's a compliance and reputation event waiting for a Friday afternoon.

Effective challenge—one of the most important ideas in MRM—needs a personality upgrade for AI. "Challenge" can't mean "recompute the AUC and call it a day." It means asking questions that feel impolite in a product meeting:

- If this model is so good, what does it get wrong *on purpose* (trade-offs), and who signed up for that?
- Where does performance come from—stable borrower fundamentals, or a fragile proxy that might vanish when the economy sneezes?
- Which segments get worse outcomes, and is that a data artifact, a policy choice, or a disguised form of prohibited discrimination?
- What happens when the input distribution shifts—when unemployment rises, stimulus ends, fraud patterns mutate, or an underwriting policy changes?

Those questions don't attack the team; they protect it. A model that survives uncomfortable questions is the one you want running at scale.

Challenger models are one of the cleanest ways to make challenge real rather than ceremonial. A challenger can be a simpler model built on the same data, a benchmark scorecard, or even a deliberately constrained ML model that trades a bit of performance for interpretability. The point isn't to "win" with the challenger; it's to expose dependence. If the champion's edge disappears when you remove a handful of features, or when you test it on a different time period, you've learned something more valuable than a leaderboard statistic: you've learned where the model is brittle.

Stability testing deserves its own spotlight because AI systems can be *quietly unstable*. Imagine a credit model that uses cash-flow volatility inferred from transactions. In a stable economy, that feature might be predictive and fair. During a disruption, the same feature can become a proxy for job sector, geography, or life events—and the model can tighten credit right when households need it most. A stable AUC can hide a shifting moral profile.

One practical way to force stability into the conversation is to test models like engineers test bridges: under load. That means out-of-time testing (train on earlier vintages, test on later ones), stress scenarios (synthetic shifts that mimic downturn behavior), and sensitivity analysis (how much does the output move when a key feature is perturbed within plausible bounds?). These tests don't predict the next crisis. They reveal whether the model fails gracefully—or fails like a glass table.

Interpretability is where MRM and banking culture often collide. Many ML models are not naturally explainable, at least not in a way that maps neatly to human language and regulatory expectations. But "black box" is not a permission slip. Banks still need to produce coherent reasons for decisions, especially adverse actions. That forces a design choice: either constrain model complexity so explanations are more faithful, or invest in explanation methods and governance that make it clear what the explanations *are* (helpful signals) and what they *aren't* (the model's inner monologue).

A useful mental model is to separate *global* understanding from *local* explanations. Global understanding answers: "What generally drives the model?" Local explanations answer: "Why did this applicant get this outcome?" MRM should require both, and it should require them to be tested. If an explanation method is unstable—if tiny changes in inputs flip the stated "top reason"—that's not just a model problem. It's a customer experience problem. People will accept bad news faster than they accept nonsense.

Fair lending and bias testing can't be stapled on at the end like a compliance garnish. AI can find predictive power in variables that are correlated with protected characteristics, even if the model never sees race or gender. In practice, that means the second line should test for disparate impact across relevant segments, examine feature candidates for proxy risk, and evaluate whether performance gains are worth the legal and reputational exposure. The bank is not only optimizing an objective function; it is defending a social license to operate.

The uncomfortable truth is that many banks don't have "a model." They have a model *embedded in a system*. Data flows in from vendors, internal platforms, and customer channels; features are computed; decisions are combined with policy rules; thresholds are tuned; exceptions are granted; adverse action reasons are generated; monitoring alerts are triaged. Model risk is often system risk wearing a model's name tag. MRM has to widen its lens to include data lineage (where the data came from), operational controls (who can change what), and downstream usage (how humans and systems act on the output).

That's why "use risk" matters as much as "build risk." A strong model can be used badly: applied to a population it wasn't trained on, run with a threshold that doesn't match the bank's risk appetite, or plugged into a workflow where staff treat the score as a command rather than an input. MRM should require a clear statement of intended use and explicit *prohibited* uses. The fastest way for AI to become a scandal is for an internal user to discover a clever, unofficial shortcut.

The first line of defense—model owners, data scientists, product teams—should be held to a standard that feels more like manufacturing than research. Reproducibility is not academic purity; it is operational safety. If the team can't reproduce last month's model from version-controlled code and a defined data snapshot, then the bank is one incident away from arguing with itself about what happened. That argument is expensive even when you win.

The second line—independent validation—needs both technical depth and institutional courage. Technical depth because ML requires specialized testing. Courage because the validator's job is to be unpopular at precisely the moment the business wants to launch. The validator is not there to slow things down; they are there to prevent the organization from shipping confidence it hasn't earned.

The third line—internal audit—should not attempt to out-math the data scientists. Audit wins by checking the plumbing: is the inventory complete, are the validations happening on schedule, are issues tracked to remediation, are overrides controlled, is monitoring effective, are approvals documented, are vendor dependencies understood, and are changes governed? In a well-run bank, audit doesn't "catch" people; it catches the gaps that let well-intentioned people drift into unsafe behavior.

A compact way to see how AI reshapes traditional MRM is to compare what used to be sufficient with what is now necessary:

MRM Element	*Traditional approach*	*AI-era adaptation*
Model definition	"Scorecards and pricing models"	Any algorithmic system influencing decisions, including feature pipelines and decision engines
Validation focus	Performance + conceptual soundness	Performance *plus* stability, drift sensitivity, fairness, explanation quality, and system integration
Documentation	Technical memo + assumptions	Model card style: intended use, limits, training data snapshot, monitoring plan, explanation method, known failure modes
Change control	Periodic updates	Continuous updates: versioning, approvals, rollback plans, and traceability for data + code + parameters
Benchmarking	Compare to prior model	Challenger models to expose brittleness; simple baselines to detect "complexity with no payoff"
Monitoring	Back-testing, periodic reports	Live monitoring of data drift, performance drift, segment stability, and alert triage with accountable owners

Monitoring is where MRM stops being a governance doctrine and becomes a living habit. The moment an AI model goes live, it starts negotiating with reality. Fraudsters adapt. Customers change behavior. Marketing campaigns alter the mix of applicants. Economic conditions shift the meaning of "risk" underneath the model like a slowly moving rug. If the bank treats monitoring as a monthly report, it's essentially agreeing to learn about failures on a delayed schedule.

Good monitoring is not a wall of metrics. It is a set of *tripwires* tied to action: thresholds that trigger investigation, playbooks that define responses, and clear ownership so alerts don't die in a shared inbox. The model owner should know what happens if drift is detected: does the bank tighten policy rules, retrain, roll back, or temporarily route marginal decisions to a human queue? MRM should insist that these options exist *before* the model needs them.

One of the most bank-native guardrails is limits. Markets desks have limits because humans under pressure make optimistic decisions. AI deserves the same suspicion. Limits can be technical (confidence thresholds below which decisions are routed for review), operational (caps on auto-approvals in a new segment), or temporal (a "sunset date" requiring revalidation). Limits are how a bank admits a simple truth: the model might be right, but the bank should not bet the whole day on it.

Another underappreciated guardrail is override governance. Overrides are not a failure; they are a release valve. The failure is when overrides become invisible, untracked, or political. If relationship managers override declines for certain customer types, the model will look "too conservative," and pressure will build to loosen it—until the bank accidentally trains itself into worse risk outcomes. Overrides should be logged with reasons, reviewed for patterns, and fed back into policy and model design. In AI systems, human behavior is part of the data-generating process.

The most mature AI risk programs treat MRM as a contract between speed and safety. The bank can move quickly *because* the guardrails are strong: clear tiering, disciplined documentation, independent challenge, robust monitoring, and an audit trail that can survive scrutiny. Without those, speed is just another word for accumulating hidden debt—debt that comes due when the economy turns, when regulators ask questions, or when a customer's complaint lands on the desk of someone who doesn't care how impressive the model looked in a sprint demo.

The goal is not to turn AI into something deterministic. It isn't. The goal is to turn AI into something *governable*. A bank can live with uncertainty; banking was built to price it. What it can't live with is uncertainty that has no owner, no controls, and no receipts.

6.2 Navigating the Rules: Regulation and Vendor Risk

The fastest way to misunderstand "AI regulation" is to picture a new rulebook landing on a desk with a neat definition of what's allowed. Banking never works that way. What actually happens is messier and more consequential: regulators, courts, and customers converge on a single expectation—*if an algorithm helps decide who gets money, you must be able to defend the decision*. Not "the vendor can defend it." Not "the model is proprietary." *You* can defend it.

That's why outsourcing AI can feel like renting a high-performance car and discovering the speeding tickets still come to your home address. A fintech can sell you software. A cloud provider can host it. A model API can score applications at industrial speed. None of them can absorb the bank's duty to operate safely and comply with consumer protection, fair lending, privacy, and operational resilience expectations. The liability trail is boring to read and thrilling to litigate.

Two forces are tightening the screws at the same time. The first is formal regulation—especially in the European Union, where the EU AI Act turns "responsible AI" from a values statement into enforceable obligations. The second is the regulatory posture in the United States and other markets, where agencies may not have an "AI law" but do have something banks can't ignore: long-standing rules for credit decisions, adverse action notices, unfair or deceptive practices, and third-party risk. AI doesn't replace those obligations; it stress-tests them.

The EU AI Act is a good example of how the world is now thinking about AI in finance: as a *risk tier* problem, not a "technology" problem. The Act categorizes AI uses into buckets-some unacceptable, some high-risk, some limited, some minimal-and then attaches real requirements to the higher-risk tier. Creditworthiness assessment and credit scoring for consumers have been treated as high-risk use cases in the EU framework. The practical meaning for a bank is simple: if you use AI to help decide whether someone gets credit (or on what terms), you should assume you're operating in the regulatory equivalent of a school zone.

The timing matters, too, because banks don't implement compliance in a weekend. The AI Act entered into force on August 1, 2024, with obligations phasing in—early rules like prohibited practices and AI literacy requirements applying from February 2, 2025, and additional obligations (including around general-purpose AI) applying from August 2, 2025, with the bulk of requirements becoming fully applicable on August 2, 2026. That clock pushes banks toward a new habit: treating "AI readiness" less like a pilot program and more like a multi-year compliance build, with budgets, owners, and audit trails.

Here's the part that catches executives off guard: the AI Act isn't only about models you build. It's also about models you *deploy*. The Act distinguishes roles such as "providers" (those who develop and place AI systems on the market) and "deployers" (those who use them). A bank can be a deployer of a vendor system and still carry significant obligations. The bank can also become something closer to a provider if it materially modifies a system, integrates it deeply, or repackages it into a customer-facing decisioning product. In other words, the moment you customize a vendor tool, you may be customizing your legal exposure along with it.

Now put the EU lens next to the U.S. reality. The United States doesn't have a single, unified AI law for banks, but banks operate under a dense canopy of expectations: safety and soundness, consumer compliance, fair lending, privacy, and model risk governance. One of the sharpest pressure points is adverse action. If a consumer is denied credit (or offered less favorable terms), U.S. rules generally require the creditor to provide specific reasons. The uncomfortable implication for complex algorithms is straightforward: "black box" opacity is not an excuse. If a bank cannot produce specific, accurate reasons because it cannot understand what the model is doing, then the bank is effectively choosing a model it cannot lawfully operate.

Vendor pitches often gloss over that by conflating *explanation* with *interpretability*. Vendors may offer a dashboard with feature importance plots, SHAP values, or reason codes. That can be useful, but compliance needs more than a visual. It needs operational reliability: the reasons must map to the bank's actual decision factors, remain stable enough to be defensible, and be supported by documentation that doesn't evaporate when the vendor's product manager changes priorities.

A useful way to think about "regulation and vendor risk" is as two overlapping maps. Regulation tells you what outcomes and controls are expected. Vendor risk tells you whether you can actually achieve those outcomes and controls with someone else's technology in the middle. The overlap is where most AI programs either mature—or get bruised.

Third-party risk management in banking is often taught like a lifecycle: plan, perform due diligence, contract, monitor, exit. That lifecycle becomes dramatically more important with AI because the technology is more adaptive, more data-hungry, and more intertwined with core decisions. In plain terms, you're not outsourcing a static tool; you're outsourcing a moving part of your decision-making nervous system.

Start with planning, because that's where banks either keep control or quietly surrender it. Planning should answer: what function is being outsourced, how critical is it, what data will leave the bank, what happens if the vendor fails, and what the bank must be able to prove to regulators. The "must be able to prove" list is often longer than the "must be able to do" list. Proving requires evidence: logs, approvals, change records, testing results, monitoring alerts, and documented issue remediation.

Due diligence then becomes less about glossy SOC reports and more about *how the model behaves under pressure*. A vendor can have excellent security controls and still deliver a model that collapses when the economy shifts or when the bank's applicant mix changes. The bank should probe for things like training data provenance, update frequency, performance stability across segments, and what monitoring is available (and at what cost). If the vendor can't answer basic questions about drift-when the model's inputs or performance change over time-then the bank is buying a sophisticated device with no fuel gauge.

Contracting is where "you can't outsource regulatory risk" stops being a slogan and becomes clauses, schedules, and teeth. AI contracts should be treated as risk instruments, not procurement paperwork. Audit rights matter, but so do the details: audit rights to whom (the bank, regulators, independent auditors), for what (model documentation, testing evidence, change logs), and with what timing (reasonable notice is not helpful if an issue unfolds in days). Service level agreements should cover not just uptime but also response time for model incidents, access to support during regulatory exams, and notification obligations when the vendor changes a model, feature pipeline, or subprocessor.

One of the most dangerous misunderstandings in AI outsourcing is the idea that the bank is buying a "model." Often the bank is buying a *system*: data ingestion, feature engineering, model scoring, decision rules, explanation layer, and monitoring. If any of those layers are opaque, the bank's ability to validate and defend the outcome is compromised. A bank doesn't need to own the source code to manage the risk, but it does need enough transparency to perform effective testing and to understand limitations. The bank should insist on what might be called *validation-grade transparency*: the information required to independently assess whether the system is fit for purpose.

Ongoing monitoring is where vendor relationships get real. Vendors love onboarding. Monitoring is where the relationship either becomes professional or becomes wishful thinking. For AI, monitoring should include at least four categories: data drift (inputs changing), performance drift (outcomes changing), compliance drift (explanations or reason codes no longer mapping cleanly), and operational drift (changes in uptime, latency, or incident response). A vendor that treats monitoring as a premium add-on is effectively charging extra for safety.

Exit planning is the part everyone wants to skip until they can't. AI vendors can become deeply embedded, especially when models are delivered "as a service" through APIs. If the vendor relationship ends abruptly-pricing shocks, acquisition, outage, regulatory findings-the bank needs a credible transition plan. That might mean maintaining a fallback model internally, negotiating access to historical scoring outputs and decision logs, or ensuring data portability so the bank can rebuild. Without an exit plan, "vendor lock-in" becomes "decision lock-in," and that is not a technology inconvenience; it's a strategic vulnerability.

There's also a subtle but growing systemic issue: concentration risk. If many banks buy similar models from the same small set of vendors, you get correlated behavior. Correlated behavior in banking is how small problems become large ones. A shared vendor feature pipeline that misinterprets a data source change can nudge credit decisions across institutions in the same direction at the same time. It's the algorithmic cousin of everyone holding the same trade.

The human temptation is to treat vendor selection as a binary "buy vs. build" decision. In reality, it's usually "buy *and* build," and the question is which pieces you are willing to own. You might buy a feature store or a model monitoring platform, but you still need internal capability to test, challenge, and explain. You might buy a credit decision engine, but you still need internal policy governance around thresholds, overrides, and exception handling. Outsourcing should reduce operational burden, not remove accountability muscle.

A practical way to keep that muscle is to define non-negotiables-capabilities the bank will not outsource even if it outsources the tooling. Non-negotiables typically include: independent validation, model inventory governance, adverse action reason governance, monitoring and incident management ownership, and the ability to reconstruct decisions after the fact. If a vendor relationship makes any of those impossible, the bank is not outsourcing technology; it's outsourcing control.

One of the most common traps is confusing vendor "certification" with bank "compliance." A vendor may claim alignment with various standards, provide white papers, and show impressive benchmark results. None of that answers the bank examiner's real question: *how do you know this works for your portfolio, your policies, your customers, and your risk appetite*? General performance is marketing; portfolio performance is risk management.

It helps to turn vendor evaluation into a set of pointed questions that feel closer to underwriting than procurement. A bank would never approve a borrower with "trust me" as a repayment plan. It shouldn't approve a decisioning model with "proprietary" as a governance plan.

Question banks should ask	*Why it matters*
What training data sources were used, and what is excluded?	Hidden exclusions can create blind spots; hidden inclusions can create privacy and proxy-risk problems.
How often do you update the model, and how are changes communicated?	Frequent updates without controlled change notices can break validation and explanation consistency.
Can we reproduce scoring for a past decision (inputs, version, outputs, reasons)?	Regulators and complaints demand reconstruction; "we don't store that" is an unacceptable answer.
What monitoring do you provide for drift, bias, and explanation stability?	If drift is invisible, the bank learns about failure from losses or headlines.
What are the sub-vendors/subprocessors, and what happens if they change?	AI stacks are supply chains; risk flows through the chain.
What audit rights do we have, and what evidence will you provide during exams?	Audit rights without evidence delivery become ceremonial.
What is the exit plan-data portability, model fallback, transition support?	Lock-in becomes critical when the service is embedded in credit decisions.

Privacy and data governance add another layer of complexity because AI systems tend to be hungry. Vendors may request alternative data, device data, behavioral signals, or transaction-level detail to boost predictive performance. Each additional data type can improve model lift while also increasing compliance exposure and reputational risk. Customers may tolerate "bank uses my credit report." They are far less forgiving of "bank denies my loan because of something it inferred from my phone behavior," even if the inference is statistically valid. The risk isn't only legal; it's trust.

Generative AI has introduced yet another vendor risk pattern: banks consuming "general-purpose" models through APIs. These models may not be making credit decisions directly, but they can influence decisions by drafting adverse action narratives, summarizing application notes, recommending next-best actions, or triaging fraud alerts. That creates a different kind of governance question: if a model is not the official decision engine but shapes the human's decision, does it need the same controls? In practice, banks are learning that influence is a form of power, and power is a form of risk.

Contracting for generative AI and model APIs also raises operational resilience concerns. If your customer service assistant or internal analyst tool depends on an external model endpoint, outages and quality regressions become a front-line experience problem. Latency is no longer an IT metric; it becomes a customer abandonment metric. A model provider's change in behavior can become your brand's change in behavior overnight. The bank should treat external model dependencies the way it treats any critical third-party service: clear SLAs, incident communication, and the ability to degrade gracefully.

A final point that tends to land hardest with senior leaders: regulatory risk is not just about fines. It's about *forced redesign*. If a regulator concludes that a bank cannot explain its credit decisions, or that a vendor model produces unacceptable bias, the remediation is rarely "tune a parameter." It can mean pausing automated decisions, reworking the model pipeline, redesigning reason codes, retraining staff, re-contacting customers, and revalidating across the portfolio. The cost is operational, reputational, and strategic—and it tends to arrive at the worst possible time, when the bank is already dealing with elevated credit losses or market volatility.

The banks that navigate this environment well don't treat regulation as a tax on innovation. They treat it as a design constraint that forces clarity: who owns the decision, what evidence supports it, how the bank monitors it, and how the bank exits it. Vendor relationships then become accelerators rather than liabilities—because the bank has already decided what it will never outsource: responsibility.

6.3 Keeping the Lights On: MLOps and Continuous Monitoring

A credit model that performs beautifully on a laptop can fail in production for the most humiliating reason imaginable: the world moved. Not in a dramatic, headline way—just a quiet shift in customer behavior, a new fraud tactic, a policy tweak, a data vendor changing a field definition. Machine learning has a talent for turning those small movements into large surprises, because it learns patterns that are partly economic truth and partly accidental choreography.

Banks already understand this dynamic in another form. A portfolio's "expected loss" changes when unemployment rises or when a new marketing channel brings in a different mix of borrowers. That's normal risk. MLOps—Machine Learning Operations—is what happens when you accept that a model is also a portfolio: it has a life, it accumulates exposure, and it requires active management. The difference is that a model's risk can compound invisibly, inside pipelines and dashboards, until the first sign of trouble is a spike in charge-offs or a regulator asking why your approvals suddenly shifted.

MLOps sounds like an engineering term because it is one, but its job is surprisingly human: to prevent organizations from lying to themselves. Without disciplined operations, teams remember the model they *meant* to deploy, not the one actually running at 2:00 a.m. They remember the training data they *think* they used, not the messy subset that ended up in the job that succeeded after three failures. They remember the monitoring they *planned*, not the alert that fired every day and got muted. MLOps makes the truth hard to avoid.

Production machine learning carries a special kind of "technical debt"—future pain caused by shortcuts taken today. The debt doesn't look like a broken system; it looks like a system that is increasingly expensive to change safely. Every new feature adds a dependency. Every dependency adds the chance that the model will behave differently tomorrow without anyone touching the code. In consumer lending and fraud, where data comes from many sources and decisions feed back into the next set of applications, technical debt is not theoretical. It's the reason "simple updates" become multi-month projects with cross-functional war rooms.

A bank that takes MLOps seriously ends up with a comforting property: the ability to be wrong *quickly* and recover *cleanly*. That sounds odd until you've lived through a model incident. The worst incidents are not the ones where a model drifts; drift is inevitable. The worst incidents are the ones where no one can answer three basic questions: what changed, when did it change, and how do we roll back?

The operational story of machine learning begins with a deceptively plain word: *data*. In practice, "data" is not a thing; it's a supply chain. It has upstream producers (core systems, bureaus, open banking feeds, fraud consortium signals), transformations (joins, imputation, encoding), and consumers (models, decision rules, explanations, dashboards). Data can be late, missing, duplicated, or subtly redefined. A field named "income" can change from "stated annual income" to "verified monthly income" and still arrive with the same column name, like a mislabeled jar in a kitchen. If you wouldn't accept mislabeled jars in a commercial restaurant, don't accept them in an automated credit decision.

A useful way to keep the supply chain honest is to baseline it. A baseline is a saved snapshot of what "normal" looks like: the distribution of key features, the rate of missing values, the mix of channels, the segment proportions. Baselines are not about perfection; they're about early warning. When a feature's distribution shifts, you want to know before your approval rate shifts. When missing values jump, you want to know before explanations degrade into nonsense like "insufficient information" for half your applicants.

Model building is the stage where many organizations still behave like scientists in a lab-iterate quickly, try new things, chase performance. That's fine, as long as the bank remembers it is also building a regulated decision system. A "good" build process isn't just a notebook that gets a high AUC. It is a pipeline that can be rerun, audited, and compared across versions without argument. It captures code, configuration, training data references, hyperparameters, and evaluation results in a form that survives the departure of the engineer who wrote it.

This is where model registries earn their keep. A model registry is basically a controlled shelf: it stores versions, metadata, and promotion states like "staging" and "production." The shelf matters because it turns model deployment from a heroic act into a repeatable process. When a new model is proposed, it doesn't arrive as an email attachment or a container image with a friendly name. It arrives as a versioned artifact with lineage, metrics, and approvals attached. That is a governance win disguised as an engineering convenience.

Deployment is where the bank pays for every ambiguous decision it made earlier. If the model expects a feature computed one way and production computes it another way, you get "training-serving skew"—a mismatch between the world the model learned and the world it now inhabits. Skew is the model equivalent of giving a chef a recipe and then silently swapping salt for sugar. The dish will still come out of the oven. That's the problem.

The temptation in banking is to treat deployment as a single moment: go-live, email, cake. MLOps treats deployment as a reversible state. Reversibility is not pessimism; it's maturity. It means you can roll back to the previous version, route traffic gradually (a "canary" release), and compare the new model to the old model on real production data before you hand it full control. Banks do this conceptually with policy changes all the time—pilot, measure, expand. MLOps makes that discipline automatic.

Monitoring is the part that sounds easy and is usually done badly. Lots of teams monitor accuracy. Fewer monitor the things that break accuracy before accuracy breaks. In lending, you often can't measure true performance immediately because default takes time. Fraud provides faster feedback, but fraudsters adapt, so yesterday's precision is not a promise of tomorrow's. Monitoring has to be multilayered: input health, model output health, decision health, and business outcome health.

Input health asks: are the features arriving, are they sane, and do they look like the baseline? Model output health asks: are scores shifting, are confidences collapsing, are explanations changing their character? Decision health asks: did approval rates, decline reasons, or override rates swing? Business outcome health asks: are early delinquency, charge-offs, fraud loss, complaints, or call center contacts changing in a way that suggests the model is steering the bank into a ditch.

There is a particular kind of monitoring that matters for banks and is often neglected: *segment stability*. A model can look stable overall while quietly deteriorating for a specific group—by channel, geography, product, income band, age proxy, small business sector, or any segment where the bank has both business and compliance exposure. Segment monitoring is where MLOps becomes a guardrail rather than a dashboard. It's also where many institutions discover their data is not as cleanly segmented as their org chart.

Drift—change over time—is the main villain in the MLOps story, but it has different costumes. *Data drift* is when the inputs change. *Concept drift* is when the relationship between inputs and outcomes changes. *Label drift* is when the outcome definition or measurement changes. Lending provides plenty of examples: a new hardship program changes repayment patterns, a new verification step changes who completes applications, a pricing change shifts the applicant pool, a fraud attack changes the distribution of legitimate versus malicious traffic. If you only watch accuracy, you discover drift late, after the losses have already been booked.

The monitoring-to-retraining loop is where banks either get ahead of drift or get managed by it. Retraining is not magic; it is controlled adaptation. Retrain too slowly, and the model becomes stale. Retrain too quickly, and you risk "chasing noise," embedding temporary quirks into decisioning logic. Retrain without governance, and you create an un-auditable system that changes its mind faster than the bank can explain it.

A stable approach is to treat retraining as a policy decision, not an engineering reflex. Define triggers: drift thresholds, performance deterioration, segment shifts, or business changes like new products. Define guardrails: minimum back-testing windows, fairness checks, explanation stability checks, and sign-offs. Define rollback rules: what constitutes a failed release. Retraining should feel less like "the model learned something new" and more like "the bank approved a new instrument."

The best MLOps programs also respect a simple truth: not every fix requires retraining. Sometimes the right move is operational. If a vendor feed is corrupted, retraining on corrupted data only teaches the model to accept corruption as normal. If a new fraud tactic appears, the right move might be to add a rule-based control while the model is retrained on verified labels. MLOps is not model worship; it's systems thinking.

MLOps is a loop, not a launch

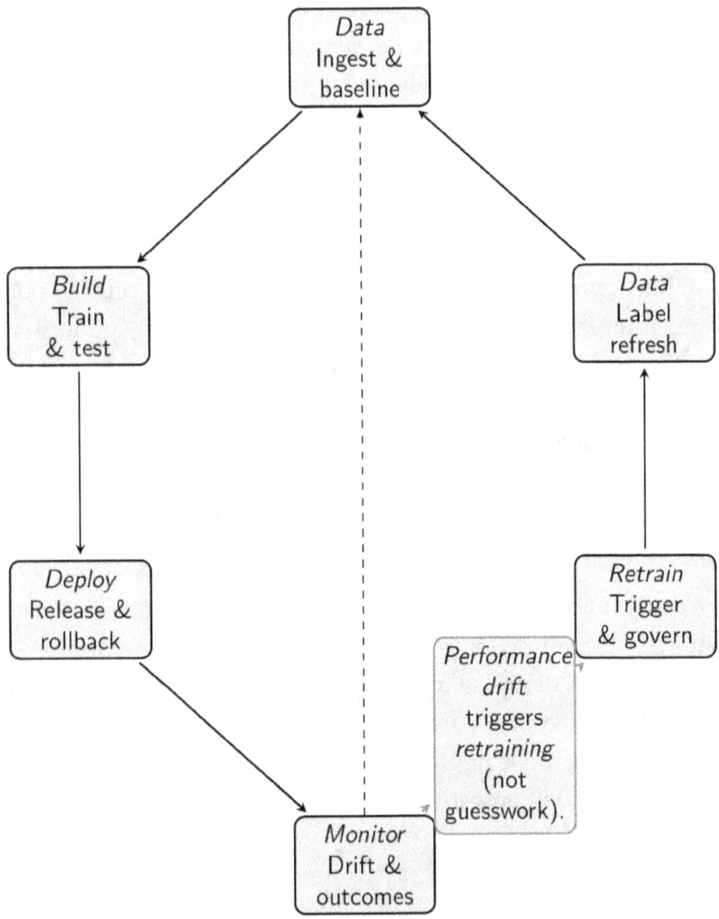

One reason MLOps feels unfamiliar in banking is that it forces a new relationship with time. Traditional credit models might be redeveloped annually. A modern ML underwriting model might need quarterly tuning, monthly calibration, or continuous monitoring that triggers retraining when drift crosses a threshold. That cadence can be uncomfortable for risk and compliance teams trained on periodic model governance. The answer isn't to slow the model down to match old processes; it's to modernize the processes to keep up without losing control.

The highest-functioning teams take a page from how banks manage other operational risks: they define incidents. A "model incident" might be a sudden change in approval rates, a spike in manual overrides, a missing data feed, an explanation layer producing nonsensical reasons, or a monitoring alert that indicates a segment is deteriorating. Once you define incidents, you can define response: who gets paged, what gets frozen, what gets rolled back, what gets escalated, and what gets documented for later review.

Documentation in MLOps should not be a static binder. It should be the exhaust trail of the system. Every deployment should produce a record: model version, code commit, feature set, training data reference, validation summary, approvals, and monitoring configuration. Every alert should have a disposition: acknowledged, investigated, false positive, mitigated, retraining triggered, or rolled back. This is not bureaucracy; it's the difference between a bank that can explain a bad week and a bank that can only apologize for it.

A bank also has to manage "hidden feedback loops," where model outputs change the data the model later learns from. Lending provides a classic example: if a model declines more applicants from a segment, the bank collects fewer repayment outcomes from that segment, which can degrade future learning and amplify bias or uncertainty. Fraud has its own version: if the model blocks certain behaviors, fraudsters move, and the label distribution shifts. MLOps has to treat these loops as part of reality, not as edge cases, because they can turn a well-intentioned model into a self-fulfilling machine.

Calibration is another underrated operational tool. Many bank decisions depend not just on ranking risk but on the numerical meaning of probabilities. If the model says 2% default probability, does that number really mean 2%? Calibration drifts over time, especially when the economic regime changes. A model can still rank-order customers well while misestimating absolute risk, leading to mispriced loans and distorted capital planning. Monitoring should include calibration checks, and remediation may involve recalibration without full retraining—an operationally lighter fix with real financial impact.

There's also a purely practical reason MLOps matters: cost control. Without automated pipelines and disciplined monitoring, banks pay for ML with human labor—expensive analysts manually pulling reports, chasing data issues, reconciling versions, and conducting ad-hoc investigations. That labor tends to show up as "overhead," which is precisely how important safety work gets squeezed. A well-designed MLOps platform shifts the cost curve: fewer heroics, more routine controls, and faster recovery when something breaks.

Vendor tools can help, but they don't remove the need for internal ownership. Monitoring platforms may offer drift detection, bias metrics, or feature attribution tracking. Model registries may provide versioning and approvals. Those tools are useful scaffolding, but the bank still has to decide what to monitor, what thresholds matter, and what actions follow. Tools measure; governance decides.

The most bank-friendly way to think about MLOps is as operational resilience for models. Just as banks plan for system outages, market shocks, and vendor failures, they must plan for model degradation. The goal is not to prevent all drift; that's like preventing all weather. The goal is to detect drift early, understand it quickly, and respond in a way that is controlled, documented, and aligned with risk appetite.

When MLOps works, it feels almost boring. Models get promoted through environments with clear approvals. Monitoring alerts are triaged and resolved. Retraining is triggered by defined criteria and produces auditable changes. Rollbacks are quick. Incidents are contained. The bank doesn't rely on memory or heroics; it relies on process. And in a world where AI systems are increasingly woven into lending decisions, fraud defenses, and customer interactions, "boring" is not an insult. It's a competitive advantage.

6.4 The Human Element: Operating Models and the Road Ahead

A bank can buy the best model in the world and still run it badly, the same way a hospital can buy an MRI machine and still misdiagnose patients. The missing ingredient is rarely compute. It is choreography: who is allowed to decide, who is obligated to challenge, and who gets woken up when the system starts behaving strangely. AI changes that choreography because it introduces a new kind of worker—one that makes recommendations at scale, never gets tired, and cannot be cross-examined in plain English unless you build the cross-examination into the job.

The most practical question an executive can ask about an AI system isn't "How accurate is it?" It's "Who is accountable for the outcome?" Accountability is not a vibe; it's an operating model. If a model denies a borrower incorrectly, triggers unfair outcomes, or green-lights a wave of synthetic-identity fraud, the post-mortem will not be a seminar on machine learning. It will be a hunt for names: who approved it, who monitored it, who overrode it, who ignored the alert, who owned the vendor relationship, who had the authority to pause the system, and who had the courage to do it.

Operating models matter because AI doesn't just automate decisions—it redistributes judgment. In a manual world, judgment sits with frontline staff, managers, and committees. In an AI world, judgment gets split: part of it is encoded into the model, part into policy rules, part into thresholds and overrides, and part into the humans who still touch the edge cases. Splitting judgment can be powerful; it can also be a loophole factory. When everyone owns a slice, it becomes easy for no one to own the whole.

Banks often talk about "human-in-the-loop" as if it were a single setting—like turning on cruise control. In reality, there are multiple ways humans can be involved, and the differences are expensive. A human can be the *decider* (the model advises), the *reviewer* (the model decides, humans check), or the *supervisor* (the model decides, humans oversee the system). Each role implies different staffing, training, tooling, and legal risk. A bank that treats these as interchangeable will discover, usually during an incident, that "in the loop" was just a comforting phrase.

The most common early-stage pattern is the "shadow model": AI runs quietly in parallel, producing scores and recommendations while humans keep making the official decisions. This can be a smart way to de-risk adoption—until the shadow becomes a ghost decision-maker. Staff begin to rely on it. Managers begin to ask why anyone is ignoring it. A tool intended for learning becomes a tool of pressure. The bank should be explicit: if AI is advisory, it stays advisory. If it becomes decisioning, governance and controls must change on purpose, not by habit.

When AI does become decisioning, the bank faces a delicate design problem: how to preserve speed without losing meaning. Automation is seductive because it removes friction. But friction is sometimes the point. In lending, the friction is where exceptions are discussed, documentation is reviewed, and borrowers are treated like humans rather than rows in a table. In fraud, friction is where a false positive becomes a customer apology—or a customer churn event. An operating model should treat friction like a scarce resource: remove it where it's wasteful, keep it where it protects customers and the bank.

A useful lens is to map where the model's output becomes *action*. Actions in a bank are not limited to "approve" or "decline." They include line assignment, pricing, documentation requirements, manual review queues, account holds, suspicious activity referrals, collections intensity, and even how call center scripts are prioritized. The moment AI influences action, it influences customer experience. And the moment it influences customer experience, it becomes brand risk. That's why operating models for AI are ultimately customer models too.

One thing AI forces into the open is the difference between *policy* and *prediction*. Policy is the bank's choice: how much risk it wants, what it will tolerate, what it will decline even if profitable, what it owes customers as fairness and dignity. Prediction is the model's estimate: likelihood of default, fraud propensity, probability of churn. When banks blur the two, models become scapegoats. "The model made us do it" is not an acceptable sentence inside a regulated institution. A clean operating model keeps policy decisions human-owned and model decisions measurable.

That separation becomes especially important with pricing. Pricing engines often combine a risk estimate with a profitability objective and constraints. If the bank can't explain which part of the price is risk, which part is policy, and which part is profit optimization, then it's not really pricing—it's improvising with math. Customers experience that improvisation as arbitrariness. Regulators experience it as a question: can you show that pricing outcomes are controlled and non-discriminatory? The operating model should make those answers available on demand, not after a scramble.

The people problem in AI is often framed as "we need more data scientists." Banks do need technical talent, but the larger constraint is connective tissue. AI systems sit across silos: business, risk, compliance, legal, operations, IT, audit, vendor management, and often marketing. If those groups interact only through tickets and escalations, the bank will build a technically impressive system with organizational blind spots. The operating model has to make cross-functional collaboration routine, not heroic.

One practical approach is to create a clear "product owner" for each high-impact AI system—someone responsible for business performance and customer outcomes—and pair them with a "risk owner" responsible for control effectiveness. That pairing should not be a tug-of-war. It should be a joint obligation to keep the system profitable *and* governable. The best pairings behave like co-pilots: one watching the destination, the other watching the instruments, both empowered to call off the flight.

Empowerment is the key word. A bank can have monitoring dashboards that detect drift, bias, or unusual overrides and still fail because no one has authority to act quickly. If pausing a model requires three committees and a monthly meeting, then the bank has effectively chosen to accept model incidents as normal. A mature operating model pre-authorizes certain actions—traffic throttling, reverting to a prior model, temporarily routing decisions to a manual queue—based on defined triggers. This is operational resilience applied to decisioning.

Resilience deserves emphasis because AI introduces new dependencies. A credit model may rely on a bureau feed, a transaction aggregator, a device fingerprint vendor, and a cloud feature store. A generative AI assistant may rely on external model endpoints. Each dependency is a failure mode. Basel-style operational resilience thinking—deliver critical services through disruption—fits AI surprisingly well. The bank's goal isn't "no outages." It's "no catastrophic loss of critical decisioning capability." That requires fallback paths: simpler models, rule-based controls, manual procedures, and practiced incident drills.

Incident drills sound like theater until the first real incident, when the theater becomes muscle memory. AI incidents are often ambiguous. Is the approval rate down because the model drifted, because marketing changed, or because the data feed is missing? Is the fraud model catching more fraud, or generating more false positives that will infuriate customers? The operating model should treat ambiguity as normal and still force action: triage, isolate, test, decide, document.

Training is another human element that gets underfunded because it looks soft. It isn't. Frontline staff need to understand what the model output *means* and what it does *not* mean. A probability is not a moral verdict. A risk score is not an instruction. Explanation codes are not always the full story, and they can be misleading if interpreted as causal truth rather than model signals. Staff who misunderstand these outputs will either over-trust the model ("computer says no") or under-trust it ("I ignore that thing"). Both are costly.

There is also a quiet psychological shift AI introduces: the erosion of dissent. Humans are social. When an algorithm provides a confident recommendation, people hesitate to argue unless the organization explicitly rewards challenge. That's why "effective challenge" cannot live only in validation teams. It needs cultural reinforcement in business operations: permission to question the model, and data to support that questioning. Otherwise, AI becomes a machine for standardizing mistakes.

Generative AI adds a twist: it can be persuasive. A credit memo drafted by a language model can sound crisp and confident even if it contains subtle errors. A customer service response can sound empathetic while being wrong. This is not the classic model risk of "the score is off." It's the risk of "the narrative is off," which can mislead both customers and internal decision-makers. Operating models for generative AI should treat outputs as drafts that require review in sensitive contexts, with clear rules about what the model is allowed to do autonomously and what must remain human-approved.

This is where socio-technical risk frameworks have become useful. NIST's AI Risk Management Framework emphasizes that AI risk is not only about performance metrics; it's about how the system interacts with people, incentives, and institutions. That framing aligns with what banks already know: behavior follows incentives. If underwriters are rewarded purely on speed, they will defer to automation. If fraud teams are rewarded purely on loss prevention, they may accept customer harm through excessive false positives. The operating model should align incentives with the bank's actual goals: profitability, fairness, resilience, and trust.

Governance is also drifting toward something that looks like a management system rather than a collection of policies. Standards such as ISO/IEC 42001 point toward AI governance that can be audited like quality management: Plan-Do-Check-Act. Banks already do this in other domains—model risk, information security, operational risk. The interesting implication is that AI governance may become less bespoke over time and more like a standardized control environment that auditors and supervisors can assess consistently. That favors banks that invest early in repeatable processes and evidence trails.

The "road ahead" question that keeps executives awake isn't whether models will get better. They will. The question is how much autonomy banks will tolerate in core decisions—and what kind of human role remains when the machine handles the routine cases flawlessly. In the near term, autonomy tends to expand in narrow slices: straight-through processing for low-risk renewals, automated fraud holds for high-confidence cases, automated outreach for collections at the right moment. Each slice feels harmless. The cumulative effect is profound: humans stop practicing certain judgments, and the organization's skill shifts from decision-making to system supervision.

That shift can be healthy if the bank invests in what supervision actually requires. Supervising an AI system is not the same as supervising people. People can explain themselves. Models need instrumentation. Supervisors need visibility into drift, segment behavior, override patterns, complaints, and the economics of errors. They need the authority to intervene, and the humility to do so when the metrics are ambiguous. In other words, they need to become a new kind of banker: part risk manager, part operations lead, part investigator.

Autonomous banking, in its fully automated fantasy form, is less interesting than the hybrid reality that is arriving faster: banks where algorithms propose, decide, and communicate—while humans set boundaries, audit outcomes, manage exceptions, and absorb accountability. The future won't be a bank without humans. It will be a bank where the humans who matter most are the ones who can keep complex systems honest.

The final human element is the one that doesn't appear on org charts: courage. AI systems will occasionally recommend actions that are profitable and uncomfortable—tightening credit in a vulnerable segment, pushing collections harder, blocking more transactions. They will also occasionally recommend actions that are reassuring and wrong—loosening credit right before a downturn, missing a new fraud pattern, smoothing over customer complaints with plausible but inaccurate explanations. A bank that wants the benefits of AI without the blow-ups needs an operating model that makes it normal to slow down, ask hard questions, and change course. That is not an anti-technology stance. It is what professionalism looks like when the decisions are fast, scaled, and consequential.

www.ingramcontent.com/pod-product-compliance
Lightning Source LLC
LaVergne TN
LVHW022205200926
848481LV00006B/1853

* 9 7 9 8 8 9 6 6 5 3 1 4 1 *